ACE MICROBIOLOGY:

(THE EASY GUIDE TO ACE MICROBIOLOGY)

BY: DR. HOLDEN HEMSWORTH

Copyright © 2015 by Holden Hemsworth

All rights reserved. No part of this publication may be reproduced, distributed, or transmitted in any form or by any means, including photocopying, recording, or other electronic or mechanical methods, without the prior written permission of the publisher, except in the case of brief quotations embodied in critical reviews and certain other noncommercial uses permitted by copyright law.

DISCLAIMER

Microbiology, like any field of science, is continuously changing and new information continues to be discovered. The author and publisher have reviewed all information in this book with resources believed to be reliable and accurate and have made every effort to provide information that is up to date and correct at the time of publication. Despite our best efforts we cannot guarantee that the information contained herein is complete or fully accurate due to the possibility of the discovery of contradictory information in the future and any human error on part of the author, publisher, and any other party involved in the production of this work. The author, publisher, and all other parties involved in this work disclaim all responsibility from any errors contained within this work and from any results that arise from the use of this information. Readers are encouraged to check all information in this book with institutional guidelines, other sources, and up to date information.

The information contained in this book is provided for general information purposes only and does not constitute medical, legal or other professional advice on any subject matter. The author or publisher of this book does not accept any responsibility for any loss which may arise from reliance on information contained within this book or on any associated websites or blogs.

WHY I CREATED THIS STUDY GUIDE

In this book, I try to breakdown the content covered in most introductory microbiology courses in college for easy understanding and to point out the most important subject matter that students are likely to encounter. This book is meant to be a supplemental resource to lecture notes and textbooks to boost your learning and go hand in hand with your studying!

I am committed to providing my readers with books that contain concise and accurate information and I am committed to providing them tremendous value for their time and money. I hope you find this guide extremely beneficial and informatory.

Best regards,

Dr. Holden Hemsworth

TABLE OF CONTENTS

CHAPTER 1: The Microbial World in Context 1

CHAPTER 2: Chemistry and Biochemistry Tie-In 9

CHAPTER 3: Observing Microorganisms 28

CHAPTER 4: Prokaryotic and Eukaryotic Cells 34

CHAPTER 5: Microbial Metabolism 49

CHAPTER 6: Microbial Growth 63

CHAPTER 7: Controlling Microbial Growth 71

CHAPTER 8: Microbial Genetics 77

CHAPTER 9: Biotechnology and Recombinant DNA 93

CHAPTER 10: Classification of Microbes 102

CHAPTER 11: Viruses 104

CHAPTER 12: Principles of Disease and Epidemiology 113

CHAPTER 13: Microbial Mechanisms of Pathogenicity 120

CHAPTER 14: Innate and Passive Immunity 125

CHAPTER 15: Practical Applications of Immunology 137

CHAPTER 16: The Immune System and Disorders 141

CHAPTER 17: Microbial Disease 148

CHAPTER 18: Antimicrobial Drugs 161

Chapter 1: The Microbial World in Context

Introduction to Microbes

Humans have to utilize specific magnifying optical instruments in order to be able to view microorganisms (microbes) as they are too small to be viewed by the naked eye.

Major Groups of Microorganisms

- Bacteria
    - Unicellular
    - Prokaryotic
    - Reproduce through binary fission
        - Form of asexual reproduction, one cell reproduces by dividing into two almost identical cells
    - Common morphology: cocci (spherical), bacilli (rod-shaped), vibrio (curved rods), spirilla (spiral shaped)

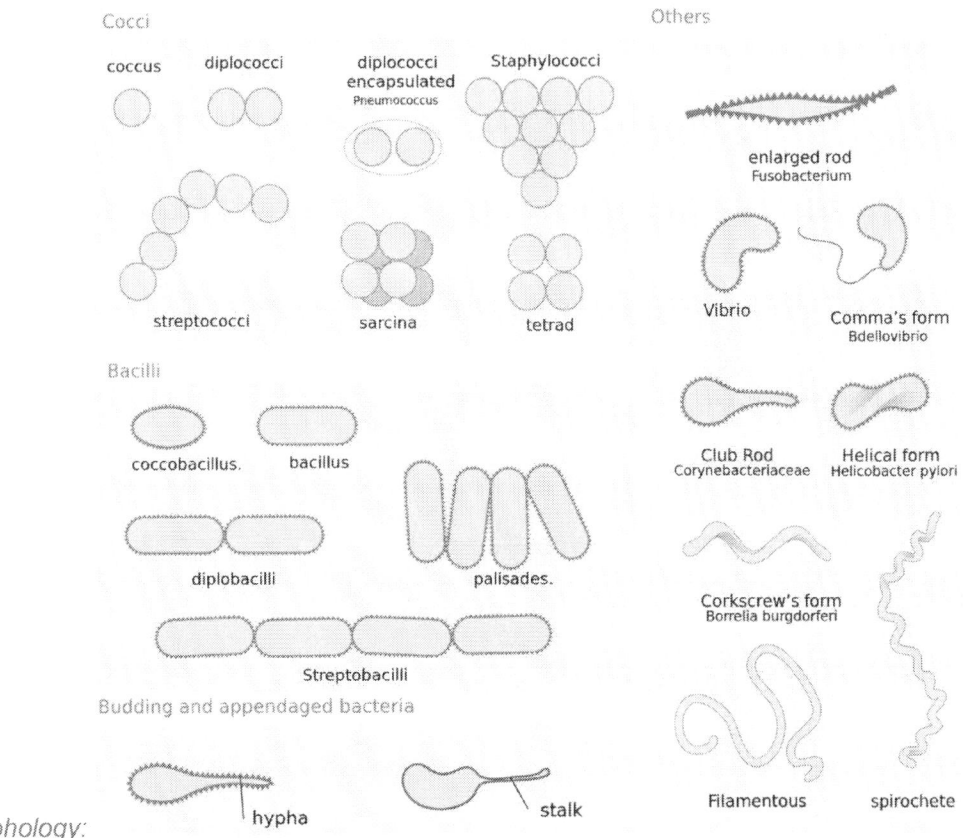

Morphology:

- Archaea
    - Unicellular
    - Prokaryotic
    - Mostly found in extreme environments
    - Three main groups:
        - Methanogens – produce methane as a waste product
        - Extreme halophiles – live in very high salt concentrations
        - Extreme thermophiles – live in incredibly high temperatures
- Viruses

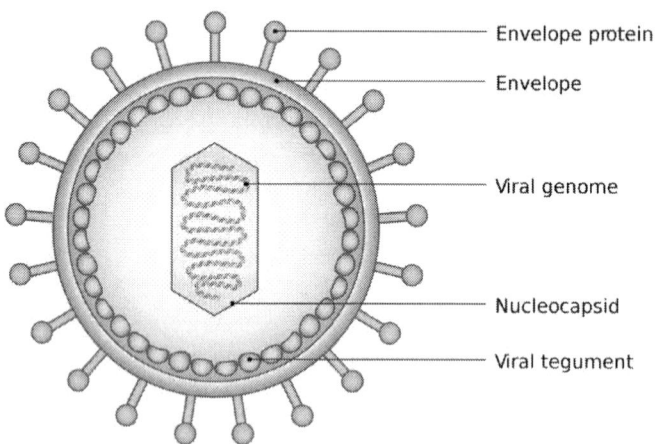

General Virus Structure Example:

    - Can only reproduce using the cellular machinery of a host cell
    - Viral genome can be either RNA or DNA
- Algae
    - Eukaryotic
    - Photosynthetic
    - Cell walls composed of cellulose
- Protozoa
    - Unicellular
    - Eukaryotic
    - Mobility achieved through flagella, cilia, or pseudopods

- Fungi
    - Eukaryotic
    - Unicellular or multicellular
    - Cell walls composed primarily of chitin

Nomenclature

The system of naming organisms was established in 1735 by Carolus Linnaeus.

- Scientific names are Latinized
- Each organism has a genus name and specific epithet (species name)
- Both the genus and species names are either underlined or italicized
- The first letter of the genus name is always capitalized
- The species name is entirely lowercase
- After being first used in its entirety, the name can be abbreviated in the text that follows by using an initial of the genus followed by the full species name

Example: *Escherichia coli* or *Escherichia coli*

- Genus Name: *Escherichia*
- Species Name: *coli*
- Abbreviation: *E. coli*

The Three Domains of Life

Introduced by Carl Woese in 1977, the three domains of life is a system of biological classification based off of rRNA sequences of organisms.

- Bacteria
- Archaea
- Eukarya
    - Includes: Protists, fungi, plants, animals

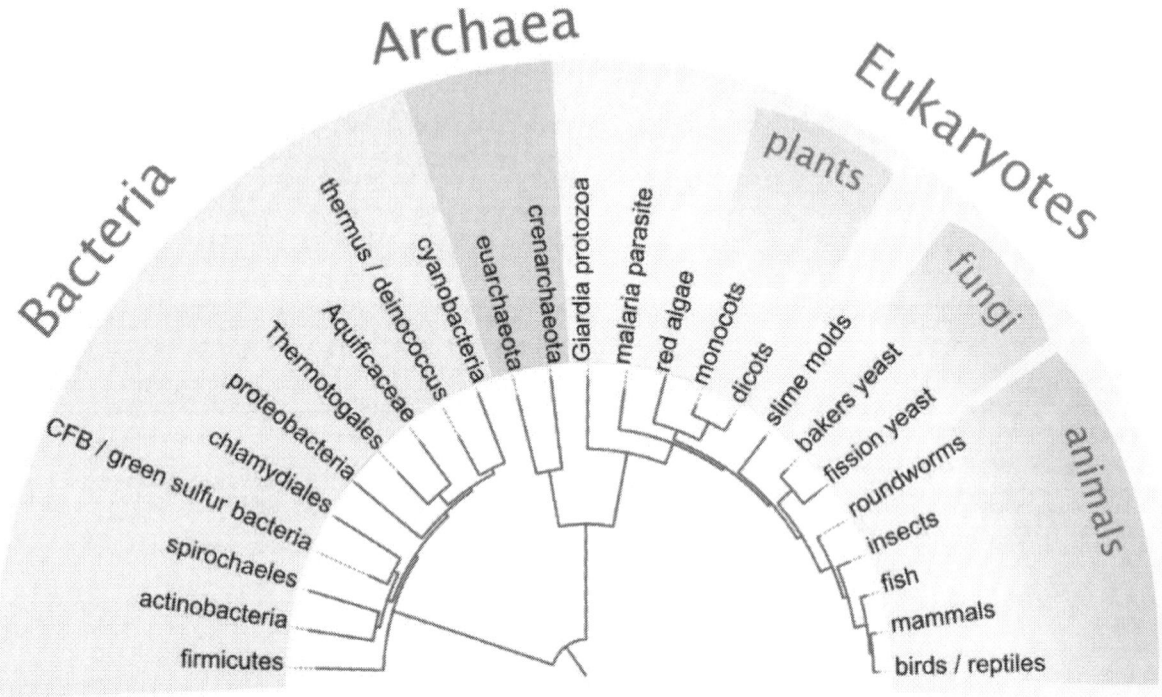

Abbreviated History of Microbiology

At the start of the development of microbiology two hypotheses sought to explain the origin of life, spontaneous generation and biogenesis. Spontaneous generation claimed that living organisms come from nonliving matter and biogenesis claimed that living organisms arise from preexisting life.

Competitive Hypotheses

- 1665 - Robert Hooke reported that living things were composed of cells
- 1668 - Francisco Redi demonstrated that maggots did not arise spontaneously from decaying meat
- 1673 - Antoni van Leeuwenhoek observed the first microbes and called them "animalcule"

- 1857 – Louis Pasteur demonstrated that microbes (yeast) are responsible for fermentation (conversion of sugar to alcohol to make beer and wine)
- 1861 - Pasteur disproved the theory of spontaneous generation
- 1876 - Robert Koch proved that a bacterium causes anthrax and comes up with Koch's postulates

Birth of Chemotherapy

Treatments for diseases that utilize chemicals is called chemotherapy. Chemotherapeutic agents can be drugs or antibiotics (chemicals produced by bacteria and fungi that inhibit or kill other microbes).

- 1910 - Paul Ehrlich developed a synthetic drug to treat syphilis
- 1928 - Alexander Fleming discovered the first antibiotic
- 1930s - Sulfonamides (basis of several groups of drugs) were synthesized
- 1940s - Penicillin was tested clinically and mass-produced

Genetics and Recombinant DNA Technology

- 1953 - Watson and Crick developed a 3D model of the structure of DNA
- 1960 - Paul Berg inserted animal DNA into bacterial DNA and the resulting bacteria produced animal protein, first example of recombinant DNA technology
- 1961 - Francois Jacob and Jacques Monod discovered the role of mRNA in protein synthesis and developed the lac operon model

Microbes in Context

A large proportion of microorganisms help maintain the balance of the environment they inhabit and are incredibly beneficial despite their ubiquitous association with disease.

- Microbial Ecology
    - Recycle nutrients, carbon, sulfur, phosphorus, etc.
    - Degrade organic matter in sewage
- Bioremediation
    - Degrade or detoxify pollutants by breaking them down into less harmful substances (e.g., oil, mercury)

- Biological Insecticides
  - Microbes that are pathogenic to insects are alternatives to chemical pesticides

Modern Biotechnology and Genetic Engineering

Biotechnology uses living systems and organisms to develop or make products (foods, chemicals, etc.).

- Bacteria and fungi are used to produce a variety of proteins and vaccines
- Missing or defective genes in human cells can be replaced with gene therapy
- Genetically modified bacteria are used to protect crops
  - Protect from insects, freezing, drought, etc.

Microbes and Human Disease

Microbes that are normally present in and on the human body are called our normal flora (or normal microbiota).

Normal Flora (Normal Microbiota)

- Usually found on skin, oral-nasal cavities, respiratory tract, digestive tract, and urogenital tract
  - Typically not found in blood, muscle, nervous tissue, or bone
- Prevent growth of pathogens
  - Compete for resources and space
  - Some secrete toxins that inhibit the growth of pathogens
    - Example of microbial antagonism
- Produce nutrients and growth factors humans can utilize
- Some normal flora may act as opportunistic pathogens
  - May cause an infection in immunocompromised individuals

Resistance

- Ability to ward off disease
- Natural resistance factors include: skin, stomach acid, immune system

Disease Terminology

- Colonization – organism does not interfere with normal physiology of the host
    - But microbe has established itself and is multiplying on/in the host
- Infection – microbe has a parasitic relationship with the host
- Disease – a disturbance in normal functioning of an organism
    - Presence of organism leads to damage to the human host
    - Damage arises directly from invading organism (e.g. toxin) or from the host immune response to the organism
- Strict/primary pathogens – organism always associated with disease
- Opportunistic pathogens – organisms that are typically members of normal microbial flora
    - But can establish disease when found in unprotected sites or in compromised hosts
- Exogenous infection – disease arising from exposure to an external source
- Endogenous infections - disease arising from host microbial flora
- Communicable (contagious) disease – spreads easily from one host to another
- Incubation period – interval of time between introduction/exposure of an organism to host and the onset of illness
    - Duration
        - Acute - Rapid onset of symptoms
        - Chronic - Infection develops slowly and lasts longer
        - Latent - Infection never completely eliminated
        - Convalescence – period of recuperation and recovery after illness
            - Infectious agent may still spread
- Infectious disease – caused by a microbe and can be transmitted from host to host
- Zoonotic disease - infectious diseases of animals that can cause disease when transmitted to humans
- Pathogens – microbes frequently associated with disease production

- Pathogenesis – mechanism a microbe uses to cause the disease state
- Infection – refers to the replication of a pathogen in or on its host
- Virulence – measure of the severity of disease a pathogen can induce

Emerging Infectious Diseases (EID)

EIDs are new or changing diseases that are increasing or have the potential to increase in incidence

- Are becoming prevalent
    - Microbial adaptation/mutation
    - Global human travel
    - Humans moving into previously uninhabited locations
- Examples
    - Bovine spongiform encephalopathy (Mad Cow Disease)
        - Caused by a prion
    - Ebola hemorrhagic fever (Ebola virus)
        - Causes fever, hemorrhaging, and blood clotting
        - Outbreaks every few years
    - Acquired immunodeficiency syndrome (AIDS)
        - Human immunodeficiency virus (HIV)
        - Sexually transmitted disease affecting males and females

CHAPTER 2: CHEMISTRY AND BIOCHEMISTRY TIE-IN

This chapter serves largely as a review of material covered in other classes. It covers most of the basic information required to understand the topics that will be covered in later chapters. Some microbiology classes may not cover this chapter in detail, as it may be expected that you already know it. Make sure you are comfortable with this information as the knowledge of this information is needed in almost every science class you will take.

Matter, Elements, and Compounds

Chemistry is the branch of science that deals with the identification and investigation of the substances that matter is composed of.

Matter

- Has mass and takes up space
    - Mass – amount of matter an object contains

Atoms

- Atom - smallest unit of matter
    - Atoms interact to form molecules
    - Atoms are composed of subatomic particles
        - Electrons – negatively charged particles
        - Protons – positively charged particles
        - Neutrons – uncharged
    - Protons and neutrons found in the nucleus
    - Electrons orbit the nucleus in an "electron cloud"
- Atomic number - number of protons in an atom of a particular element
    - For a neutral atom: number of electrons = number of protons
    - All atoms of an element have the same atomic number (same number of protons)
- Mass number = number of protons + number of neutrons

- All atoms of an elements don't have the same number of neutrons
- Molecule - combination of two or more atoms

Chemical Elements

- Element - substance that can't be broken down into other substances by chemical means
    - Trace element – element required by an organism in very small amounts
    - Each chemical element has a different number of protons in its nucleus
- Compound – substance formed from two or more chemical elements that are chemically bonded together
- Mixture - two or more elements (or compounds) mingling without any chemical bonding

Isotopes

- Isotope - atom of an element having more or fewer neutrons than the typical amount
- Radioisotope – isotope with an unstable nucleus
    - Half-life – time it takes for ½ of the atoms of a given amount of radioactive substance to disintegrate
- Biological applications of radioactive isotopes:
    - Dating fossils
        - Radio isotopes decay at a fixed rate
        - Decay rate can be used to back calculate the age of the fossil
    - Radioactive labelling and radioactive tracers
        - Easily detected even at low concentrations
        - Can be used to trace the steps of biochemical reactions
        - Trace the location of a substance through an organism
    - Diagnose or treat disease
        - Diagnose disease using a PET scan
        - Treat cancer

Atomic Orbitals

Atomic orbitals are mathematical functions that describe the wave-like behavior of an electron in a molecule, it calculates the probability of where you might find an electron (e⁻). Orbitals can combine to form hybridized orbitals needed for molecular bonding interactions.

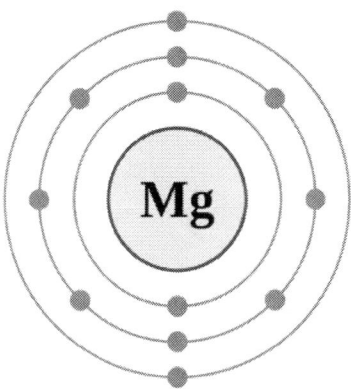

Shell Model of a Magnesium Atom

- Electrons orbit the nucleus of an atom in "orbitals" of increasing energy levels, or shells
- Orbitals aren't necessarily circular as represented in the shell model
    - In reality the orbitals are "clouds" of various shapes
- Each orbital can only hold a limited number of electrons
- Single atom can have multiple orbitals of different shapes
- Having a filled outer shell is the most stable state for an atom
- Electrons in the outer shell are called valence electrons
- Most atoms don't have filled outer shells
    - Form chemical bonds with other atoms in order to fill their outer shells

Types of Bonds

Forces holding atoms together in a compound are called chemical bonds.

Covalent Bonds

- Two atoms share valence electrons
- Indicates that atomic orbitals are overlapping
    - Overlapping requires proximity and orientation

- Two Types
  - Non-polar covalent bond – electrons shared equally between atoms
    - Electronegativity of the two atoms is about the same
    - Typically electronegativity difference between the two atoms has to be less than 0.5 for non-polar bonds
    - Electronegativity – an atom's ability to attract and hold on to electrons, represented by a number
  - Polar covalent bonds – electrons shared disproportionately between atoms
    - Electronegativity between the two atoms is different by a greater degree than 0.5 but less than 2.0

Ionic Bonds

- Electrons are transferred
  - Not shared between atoms
- Atom with high electronegativity will take an electron from an atom with low electronegativity
  - Typically, difference in electronegativity is more than 2.0
- Ion – charged atom or molecule
  - Anion – negatively charged ion
  - Cation – positively charged ion

Hydrogen Bonds

- Attractive force between a hydrogen attached to an electronegative atom of one molecule to a hydrogen attached to an electronegative atom of a different molecule
- Electronegative atom is usually an O, N, or F

Van der Waals Forces

A general term used for the attraction of intermolecular forces between molecules.

Dipole-dipole Interactions

- Interaction between 2 polar groups

London Dispersion Forces

- Interaction between 2 non-polar molecules
- Small fluctuation in electronic distribution

Intermolecular Forces

Forces that act between neighboring particles (can be repulsive or attractive).

- Intermolecular bond strength ranking (strong to weak):
    - Covalent > ionic > hydrogen > Van der Waals forces
- Weaker bonds and forces are easily broken or overcome and also re-formed
    - Makes them vital for the molecular dynamics of life

Molecular Weight and Moles

The sum of the atomic weights in a molecule is the molecular weight. One mole of a substance is its molecular weight in grams.

Example: Glucose ($C_6H_{12}O_6$)

- Molecular weights
    - C = ~12 grams
    - H = ~1 grams
    - O = ~16 grams
- Total molecular weight for glucose = (12 g x 6 atoms of Carbon) + (1 g x 12 atoms of Hydrogen) + (16 grams x 6 atoms of Oxygen) = 180 grams
- If you put 180 grams of glucose into one liter of a solution, you will have a one molar glucose solution
    - Moles = grams/molar mass
    - Molarity = moles/ volume

Chemical Reactions

Chemical reactions require energy to occur (activation energy).

- Chemical reactions involve the making or breaking of bonds between atoms
- Endergonic reactions absorb more energy than they release
    - Non-spontaneous

- Exergonic reactions release more energy than they absorb
    - Spontaneous

Synthesis Reactions

- Occur when atoms, ions, or molecules combine to form new, larger molecules
- A + B → AB

Condensation (Dehydration) Reaction

- H^+ is removed from one molecule
- OH^- is removed from another
- Form H_2O

Decomposition Reactions

- Occurs when a molecule is split into smaller molecules, ions, or atoms
- AB → A + B

Cleavage (Hydrolysis) Reaction

- Cleavage or breaking of chemical bonds by the addition of water
- H^+ and OH^- (from water) become attached to the newly exposed sites on the molecule

Ion Exchange Reactions

- Part synthesis, part decomposition
- Example: NaOH + HCl ⟷ NaCl + H2O
- Reversible reactions
    - Can readily go in either direction
        - Each direction may need special conditions

Acids, Bases, Salts, and pH

According to the Bronsted-Lowery definition an acid is proton donating and a base is proton accepting. A salt is a substance that dissociates into cations (+) and anions (-), neither of which are H⁺ or OH⁻ (e.g. NaCl → Na⁺ + Cl⁻).

- Electrolytes - substances that dissociate into ions in solution, and conduct electricity

- Amount of [H⁺] in a solution is expressed as pH

    - $pH = -\log[H^+]$ or $pH = pK + \log \frac{[A^-]}{[HA]}$

        - Increasing [H⁺] increases acidity and lowers pH
        - Increasing [OH⁻] increases alkalinity and raises pH
        - Acidic solutions contain more H⁺ than OH⁻
        - Alkaline solutions contain more OH⁻ than H⁺

    - pH scale is logarithmic
        - Each whole number pH value actually represents a 10-fold difference in concentration
    - pH 1 = most acidic, pH 7 = neutral, and pH 14 = most basic

Biological Molecules

Organic compounds contain both carbon and hydrogen atoms, while inorganic compounds typically lack carbon.

Inorganic Compound: Water

- Has 2 hydrogen atoms covalently bonded to 1 oxygen atom
- Electronic structure is tetrahedral
    - Two covalent bonds with H atoms
    - Two sets of unpaired electrons
- Polar molecule
    - Has dipoles resulting from unequal sharing of electrons
    - Polar molecules are "hydrophilic;" are attracted to water
        - Non polar molecules are "hydrophobic;" are repelled by water

- - - Dipoles can align to form H bonds

Structure of Water with Dipoles:

- H_2O can hydrogen bond with other molecules besides itself
  - Common electronegative atoms are N, O, and S
- Highly cohesive
  - Because of hydrogen-bonding interactions
  - One molecule of H_2O can hydrogen bond with 4 other water molecules
  - Hydrogen bonding between water molecules makes water a temperature buffer
    - Water can absorb a lot of heat before it becomes heated or evaporates
      - Energy is required to overcome all of the hydrogen bonds
- High surface tension

Organic Compounds

Organic compounds always contain carbons and hydrogens but they may also contain many other elements.

- Carbon can form bonds with up to four other atoms
- Carbon atoms are often found bonded together into chains and rings
- Hydrocarbons refer to compounds that are only made of carbon and hydrogen
  - Examples: methane (CH_4), ethane (C_2H_6), propane (C_3H_8)
- Chain of carbon atoms in an organic molecule is the carbon skeleton/backbone
- Functional groups are clusters of atoms that are bond to a carbon backbone
  - Functional groups are responsible for most of the chemical properties of a particular organic compound

- Small organic molecules can combine into large macromolecules
- Polymers are built from unit molecules (monomers) for structure/nutrient storage
    - Monomers in polymers are called "residues"

Carbohydrates

- Also called saccharides
    - Monosaccharides – one unit
    - Di-, tri-, etc.: ≥ 2 units
    - Larger polymers: polysaccharides
    - Can be modified with other groups to contain N, P, etc.
- Contain multiple -OH groups
- Readily convertible between open and close (cyclic) form

Carbohydrate Representation Examples

- General formula: $(CH_2O)_n$; $n \geq 3$

Fuel-storage Carbohydrates

- Starch
    - Sugar storage form in plants
- Amylose
    - Linear (unbranched) form of starch
    - Linkages create "kink" in structure
- Amylopectin
    - Same linkages as starch
        - But shorter chains

- Glycogen
    - Polymer that resembles amylopectin, but branches are about every 12 residues
    - Highly branched form allows it to be rapidly assembled or disassembled
    - Stored in muscles and liver in humans

Structural Carbohydrates

- Cellulose
    - Structural carbohydrate in plants
    - Forms extended fibers
        - Provides rigidity and strength for cell walls in plants
    - Extensive hydrogen bonding network within and between adjacent chains
- Chitin
    - Polymer of two sugars repeating many times
    - Part of the exoskeletons of insects and crustaceans
    - Part of the cell walls of many fungi

Proteins (Polymer Made of Amino Acids)

- Sometimes also called polypeptides
- Peptide (amide) bonds are formed between amino acid molecules
- Amino Acids
    - All have at least 2 ionizable groups (an amino and a carboxyl group)
    - Identities are determined by the side chain (R group)

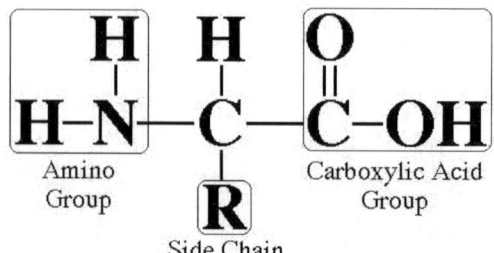

General Structure of Amino Acids

The Four Levels of Protein Structure

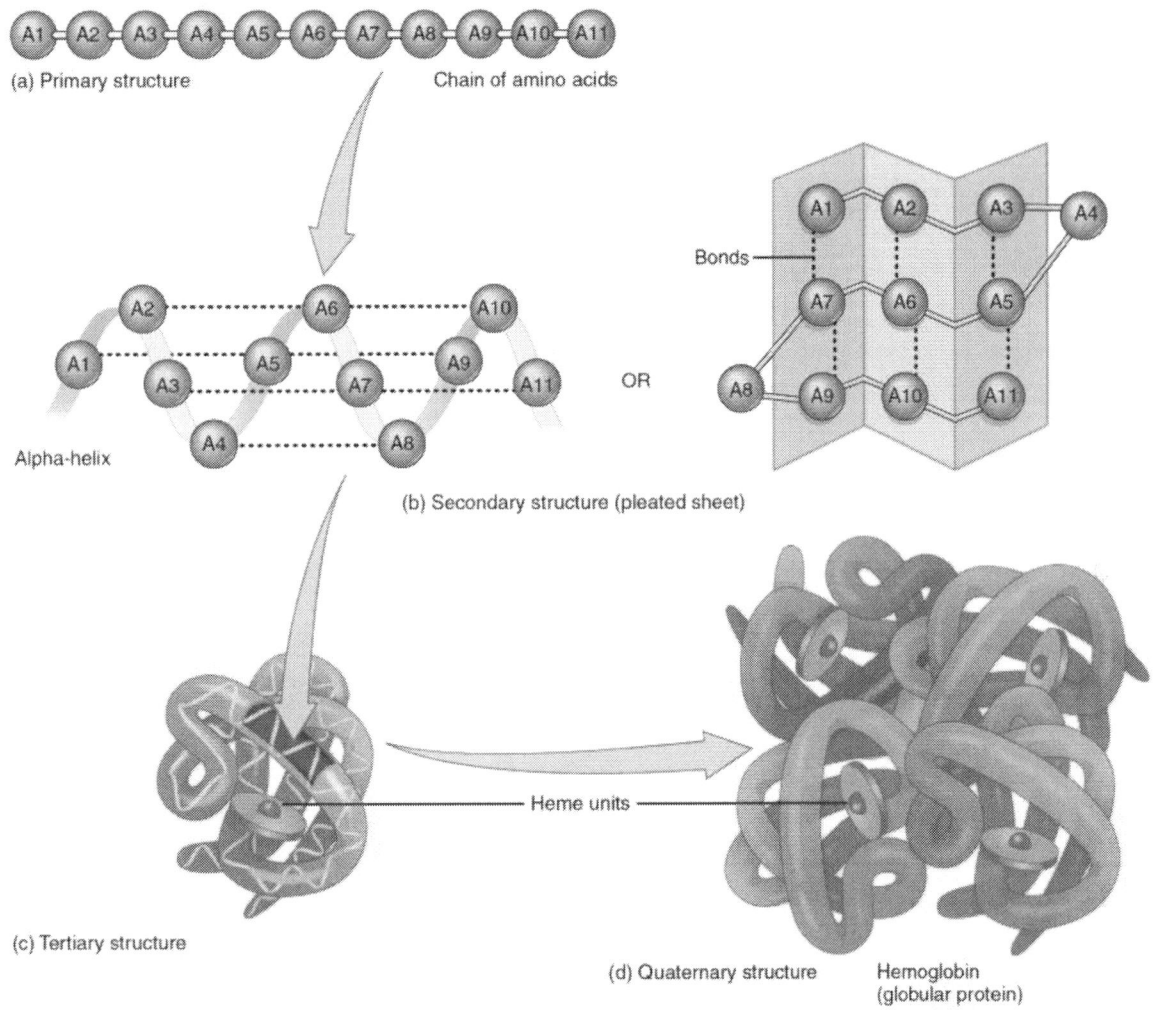

- Primary Structure
  - Sequence of Amino Acids
  - Bonds present: peptide bonds
- Secondary Structure
  - Localized conformation of the polypeptide backbone
  - Bonds present: peptide bonds
  - Two types of secondary structure:
    - Alpha Helix
      - Peptide backbone wound around a long axis core

- Forms a cylinder
- R-groups radiate outward
- 3.6 amino acids per 360° turn
- Formed by H-Bonds
                - Beta Sheet
                    - Linear zig-zag sheet of polypeptides
                    - Formed by hydrogen bonds, as well as intra and inter-chain reactions
- Tertiary Structure
    - 3D structure of an entire polypeptide
    - Bonds present: hydrogen bonds, ionic bonds, disulfide bonds, and Van der Waals forces
- Quaternary Structure
    - Spatial arrangement of polypeptide chains in a protein with **multiple subunits**
        - May have prefix homo- (identical subunits) or hetero- (different subunits)
    - Bonds present: hydrogen bonds, ionic bonds, disulfide bonds, and Van der Waals forces

Types of Proteins

- Structural proteins - provide structure (found in bone, muscle, and hair)
- Transport proteins - transport substances across cell membranes or in blood
- Regulatory proteins – hormones
- Enzymes - reduce activation energy of chemical reactions to speed up the rate of reaction
- Lipoproteins – combination of lipids and proteins
- Glycoproteins – combination of carbohydrates and proteins form when oligosaccharides are bonded to certain proteins

Lipids

- Long chain of hydrocarbons
- Mostly nonpolar
    - Generally, this makes them water-insoluble
- Can be amphipathic
    - Contain both polar and nonpolar components
- Steroids are lipids that have fused rings in their structure

Example of a Steroid, Cholesterol Structure:

Triglyceride:

Types of Lipids

- Fatty Acids
    - Long chain of carboxylic acids
    - Even-numbered chains are most common
    - Saturated fatty acids
        - Tails saturated with H atoms
        - No double bonds

- o Unsaturated fatty acids
    - Contain at least one double bond
    - Double bonds usually have *cis* configuration
- o Degree of saturation determines how they "pack"
    - Double bonds produce kinks in the chain making it harder for molecules to stack
- o Fatty acids are not typically found in free form
    - Usually esterified to glycerol
- Triacylglycerols (Triglycerides)

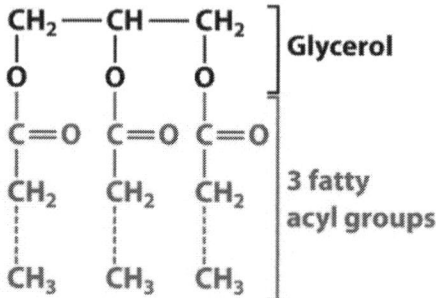

General Structure of Triacylglycerols:

- o Glycerol backbone
- o Esterified to 3 fatty acids
    - Fatty acid chains may be different
        - Provides variety of structure/function
- o Do not form bilayers (not a part of biological membranes)
- o Aggregate into globules (stored for energy)
- Glycerophospholipids
    - o Glycerol backbone
    - o 2 fatty acids + phosphate-derivative head group
        - Amphipathic because of the head group
    - o Abundant in membranes
    - o Phospholipases
        - Hydrolyze phospholipids

- May release phospho head group or fatty acid chain
- Sphingolipids
  - Sphingosine backbone
  - One acyl chain already present as part of backbone
  - Second fatty acid chain linked to sphingosine via amide bond
  - May contain a phosphate-head group
  - Head group could also have saccharide head groups
    - Cerebroside – monosaccharide head group
    - Ganglioside – oligosaccharide head group
  - Sphingomyelin - part of myelin sheath around nerves
- Sterols
  - Constructed from 5-carbon units like isoprene
  - Rigid structure

Nucleotides and Nucleic Acids

Nucleotides (Monomers of Nucleic Acids)

- 3 components of nucleotides are:
  - 5-Carbon Sugar (pentose)
  - Nitrogen containing base ring
  - PO$_3$ group (phosphate group)
    - Attached to the #5 carbon of the sugar

- Pyrimidine – nitrogenous base with a characteristic six-membered ring consisting of carbon and nitrogen atoms
    - Cytosine (C), Thymine (T), and Uracil (U) are pyrimidines

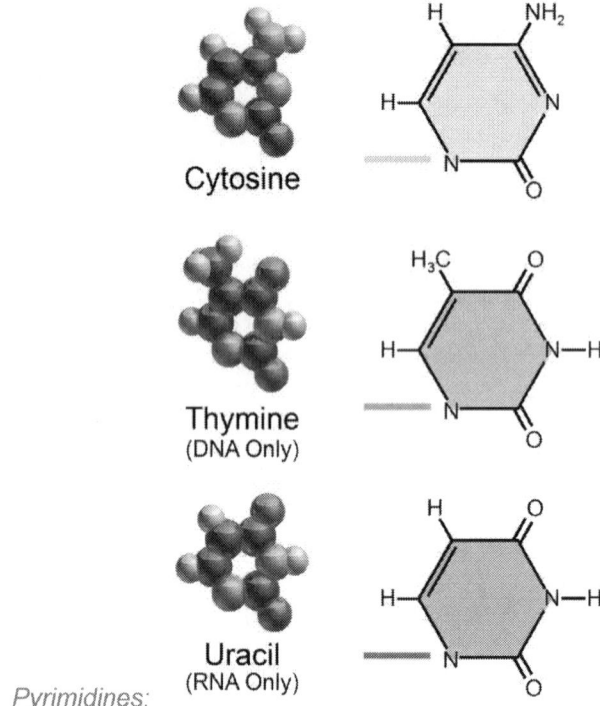

Pyrimidines:

- Purine – nitrogenous base with a characteristic five-membered ring fused to a six-membered ring
    - Adenine (A), and Guanine (G) are purines

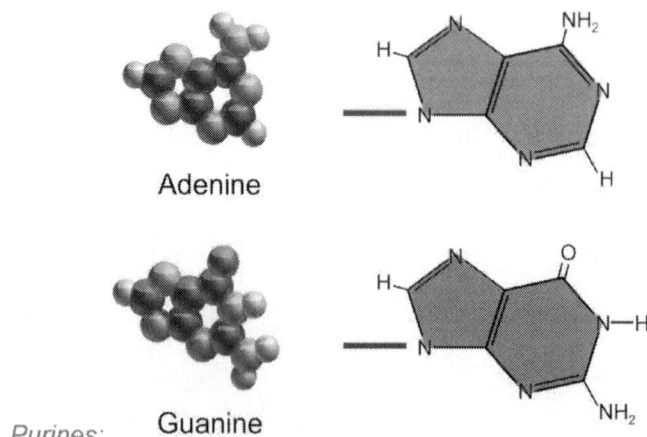

Purines:

- Phosphodiester bonds covalently link nucleotides together
    - Phosphate connects the 5' carbon of one nucleotide to the 3' carbon of another nucleotide
        - Gives strands directionality
        - In a strand all sugars are oriented in the same direction
- Phosphates and sugar form the backbone of a strand
    - Bases project from the backbone

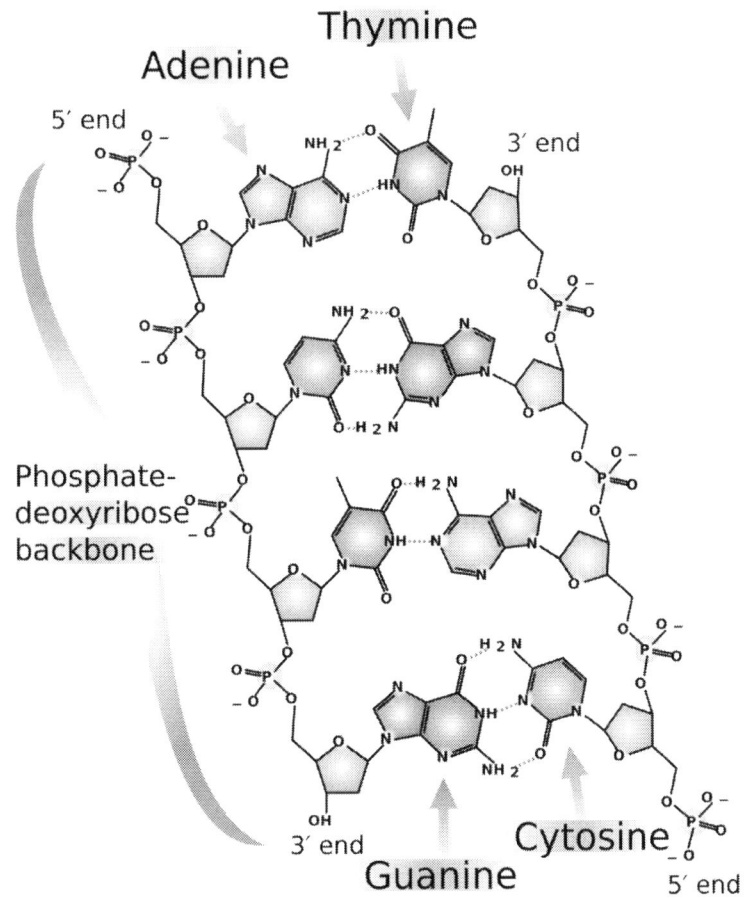

DNA Structure:

DNA (Deoxyribonucleic Acid)

- Deoxyribose as the pentose sugar
- Two nucleotide chains form a double helix

- Contains the nucleotides:
    - Thymine (T), Adenine (A), Cytosine (C), and Guanine (G)
        - A forms 2 hydrogen bonds with T
        - G forms 3 hydrogen bonds with C
- Contain genes that are instructions for protein synthesis

RNA (Ribonucleic Acid)

- Single strand nucleic acid
- Ribose as the pentose sugar
- Contains the same nucleotides as DNA except that Adenine (A) is replaced with Uracil (U)
- Functions in the synthesis of proteins
    - Messenger RNA (mRNA) carries encoded genetic message to the cytoplasm from the nucleus

RNA vs. DNA

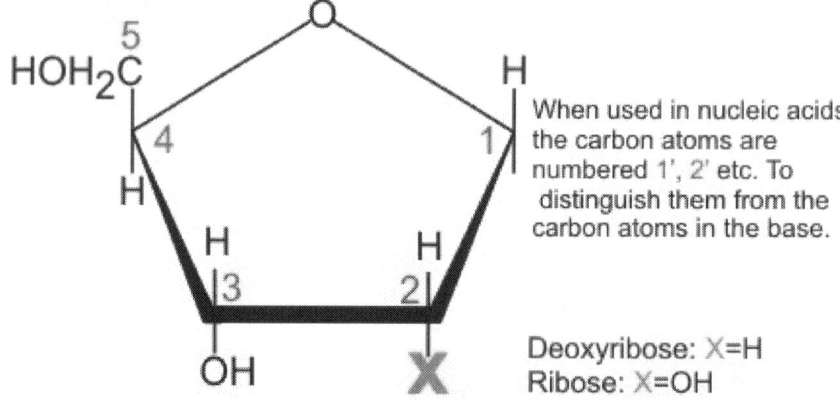

Deoxyribose & Ribose Sugars

- Position of 2'-OH of ribose prevents formation of classic Watson-Crick B helix in RNA due to steric hindrance
- 2'-O atom would come too close to 3 atoms of the adjoining phosphate & 1 atom in next base

ATP (Adenosine Triphosphate)

- ATP is a nucleotide with three phosphate groups
- ATP hydrolysis is the major way the cell has to release energy stored in chemical bonds
- Energy is required to form the bond
    - Energy is released when the bond is broken
- Takes a lot of energy to hold two phosphates together
    - Repulsion from like negative charges

CHAPTER 3: OBSERVING MICROORGANISMS

Microscopy

A simple microscope has only one lens (e.g. magnifying glass or hand lens).

Compound Light Microscopy

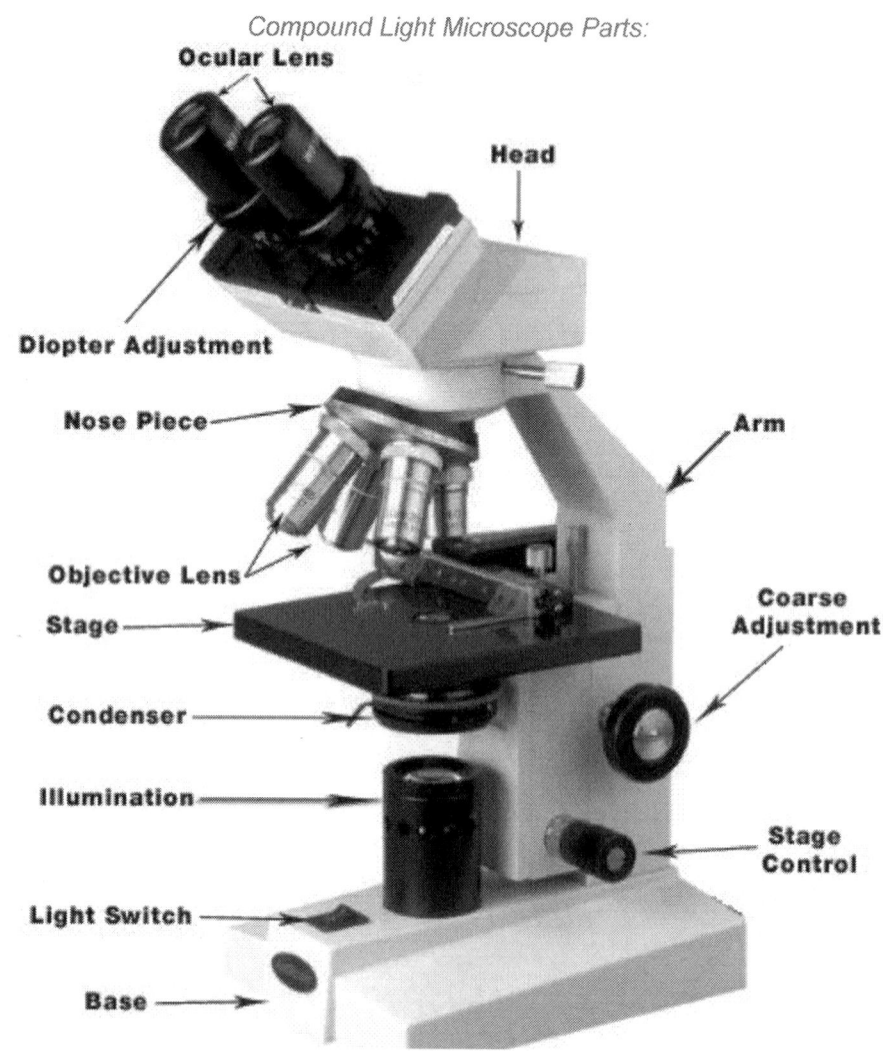

Compound Light Microscope Parts:

- Parts and Functions
    - **Eyepiece:** Lens to view the specimen
        - Usually contains a 10X or 15X power lens
    - **Diopter Adjustment:** Useful as a means to change focus on one eyepiece so as to correct for any difference in vision between the user's two eyes

Page | 28

- **Head:** Connects the eyepiece to the objective lenses
- **Arm:** Arm connects the body tube to the base of the microscope
- **Coarse adjustment:** Moves the stage to bring the specimen into general focus
- **Fine adjustment:** Fine tunes the focus by moving the stage minimally
- **Nosepiece:** Rotating turret holding the objective lenses
- **Objective lenses:** Lenses closest to the specimen
  - Standard microscope has three, four, or five objective lenses
    - Range in power from 4X to 100X
- **Stage:** Flat platform where the slide is placed
- **Stage clips:** Metal clips to hold the slide in place
- **Aperture:** Hole in the middle of the stage that allows light from the illuminator to reach the specimen
- **Illumination:** Light source
- **On/off switch:** Turns illuminator off and on
- **Iris diaphragm:** Adjusts the amount of light that reaches the specimen
- **Condenser:** Gathers and focuses light from the illuminator onto the specimen
- **Base:** Supports the microscope

- Total magnification = magnification of objective lens X magnification of ocular lens

Ocular Lens Magnification	Objective Lens Magnification	Total Magnification
10 X	4 X	40 X
10 X	10 X	100 X
10 X	40 X	400 X
10 X	100 X (oil immersion)	1000 X

- Resolution – ability of the lenses to distinguish between two points
  - Shorter wavelengths of light provide better resolution

- Refractive index - light-bending ability of a medium
  - Light may bend in air so much that it misses the small diameter of the high magnification lens
    - Immersion oil is used to keep light from bending

Brightfield Illumination

- Dark objects are visible against a bright background
- Light reflected off the specimen does not enter the objective lens

Darkfield Illumination

- Light objects are visible against a dark background
- Light reflected off the specimen enters the objective lens

Phase-Contrast Microscopy

- Converts phase shifts in light passing through a transparent specimen to brightness changes in the image

Differential Interference Contrast Microscopy

- Accentuates diffraction of the light that passes through a specimen; uses two beams of light

Fluorescence Microscopy

- Uses UV light
- Fluorescent substances absorb UV light and emit visible light
- Cells can be stained with fluorescent dyes

Confocal Microscopy

- Uses fluorochromes (fluorescent dyes) and a laser light
- Laser illuminates each plane in a specimen to produce a 3-D image

Electron Microscopy

- Uses electrons instead of light
- Shorter wavelength of electrons allows for better resolution

Transmission Electron Microscopy (TEM)

- Ultra-thin sections of specimens
- Light passes through specimen, then an electromagnetic lens, to a screen or film
- Specimens may be stained with heavy metal salts
- 10,000X-100,000X; resolution 2.5 nm

Scanning Electron Microscopy (SEM)

- An electron gun produces a beam of electrons that scans the surface of a whole specimen
- Secondary electrons emitted from the specimen produce the image
- 1000X-10,000X; resolution 20 nm

Scanned-Probe Microscopy

- Scanning tunneling microscopy uses a metal probe to scan a specimen
- Resolution 1/100 of an atom
- Produces 3-D images

Preparation of Specimens

Unstained and live cells provide little contrast with the surrounding medium which makes them harder to observe without the right preparation or equipment.

- Smear - thin film of a solution of microbes on a slide
    - A smear may be "fixed" using heat to attach the microbes to the slide

Preparing Smears for Staining

- Stains are salts that consist of a positive and negative ion
    - In a basic dye, the color is due to a cation (+)
    - In an acidic dye, the color is due to an anion (-)
- Basic dyes are attracted to bacteria because they have a slight negative charge at pH 7
- Acidic dyes typically stain the background and not the specimen
    - Staining the background instead of the cell is called negative staining

Simple Stains

- Simple stain - use of a single basic dye
- A mordant (typically an inorganic oxide), may be used to intensify the stain or increase the affinity of the stain for the specimen

Differential Stains

React differently with different kinds of bacteria, which allow them to be differentiated.

- Acid-fast staining
    - Cells that retain a basic stain in the presence of acid-alcohol are called acid-fast
    - Non–acid-fast cells lose the basic stain when rinsed with acid-alcohol
        - Usually counterstained (with a different color basic stain) to see them

Gram Stain

The Gram stain classifies bacteria into two large groups: Gram-positive and Gram-negative. Steps in making a Gram stain:

- Bacterial smear is made and heat fixed
- Smear is flooded with crystal violet for 60 seconds and then rinsed with distilled water
    - Crystal violet sticks to peptidoglycan in the cell wall of bacteria
        - Primary stain, since it colors all cells
- Smear is then flooded with iodine for 60 seconds then rinsed with distilled water
    - Iodine is a mordant
        - Increases the affinity of the crystal violet for the specimen
- Smear is then flooded with an acetone/ethanol wash for 5 seconds then rinsed with distilled water
    - Acetone/ethanol washes the crystal violet only out of the cell walls of Gram-negative bacteria
        - Since they have a much smaller amount of peptidoglycan in their walls

- o Gram-positive cell walls retain the crystal violet
  - ▪ As long as the acetone/ethanol wash is not applied for too long
- o This is the "differential step" in the Gram staining process
  - ▪ Gram-positive cells look purple after this step
  - ▪ Gram-negative cells look clear
- Smear is then flooded with safranin for 1.5-2.0 minutes then rinsed with distilled water
  - o Safranin is pink in color and stains all bacteria
    - ▪ Gram-positive cells have a purple cell wall due to the crystal violet and because the pink color is not strong enough to show through
  - o Gram-negative cells appear pink
  - o Safranin is a counterstain
    - ▪ Has a different color than the primary stain (crystal violet)
- Slide is then dried and is ready to be viewed under a microscope

Gram-positive Bacteria

They have a thick peptidoglycan layer in their cell walls which is the part that stains purple due to the crystal violet and iodine washes during Gram staining. They tend to be killed easily by penicillin, cephalosporins, and detergents which target cell walls.

Gram-negative Bacteria

They have a thin layer of peptidoglycan in their cell walls, and aren't able to hang on to the small amount of crystal violet dye that they pick up once they are flooded with the acetone/ethanol wash during Gram staining. They contain a layer of lipopolysaccharide (endotoxin) as part of their cell walls. They are more resistant to antibiotics, which can't penetrate the lipopolysaccharide layer very well but they tend to be easily killed by streptomycin, chloramphenicol, and tetracycline.

Chapter 4: Prokaryotic and Eukaryotic Cells

Size and Morphology of Bacterial Cells

Most bacteria are monomorphic; they do not change shape unless environmental conditions change. A few are pleomorphic; come in a variety of shapes.

Size

- Average size of prokaryotic cells: 0.2 -2.0 µm in diameter, 1-10 µm in length
- Typical eukaryote: 10-500 µm in length
- Typical virus: 20-1000 nm in length

Morphology

- Cocci – spherical
- Bacilli – rod shaped
- Vibrio – curved rods
- Spirilla – spiral shaped
- Spirochetes - flexible spiral shape

Cell Number Descriptive Prefixes

- Diplo - two cells
- Tetra - four cells
- Sarcinae - cube of 8 cells
- Staphylo - clusters of cells
- Strepto - chains of cells

External Prokaryotic Structures

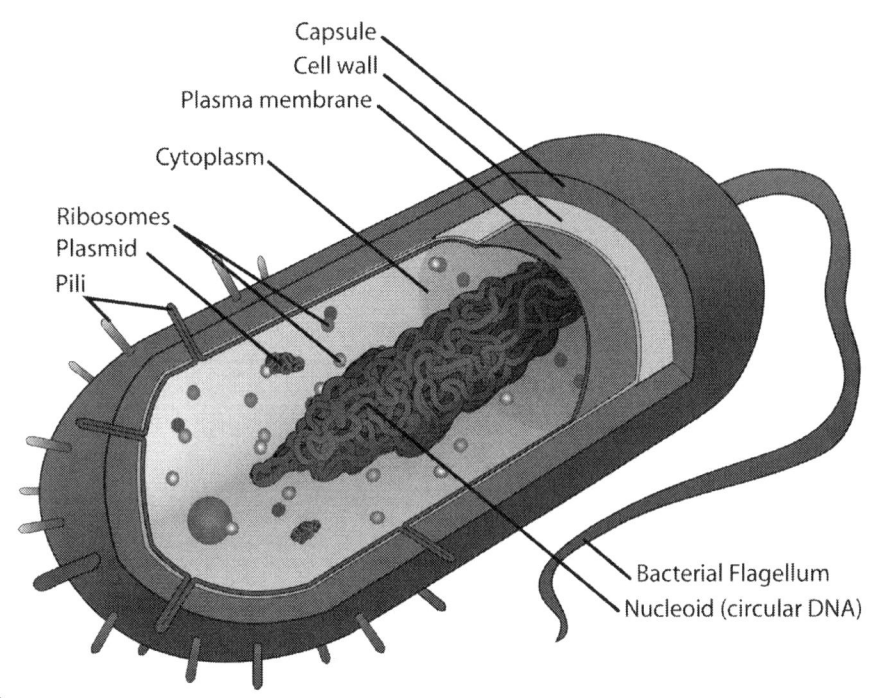

Prokaryotic Cell Diagram:

Glycocalyx

- Glycoprotein-polysaccharide cover that surrounds cell membranes of some bacteria
- Can be broken down and used as an energy source when resources are scarce
- Helps keep nutrients from moving out of the cell
- Different types
    - Capsule - neatly organized and attached to the cell wall
        - Capsules prevent phagocytosis by the host's immune system
    - Slime layer - unorganized and loosely attached to the cell wall
    - Extracellular polysaccharide - allows bacterial cells to attach to various surfaces

Prokaryotic Flagella

- Long, semi-rigid, helical, cellular appendage used for locomotion
- Consists of the protein flagellin
    - Attached to a protein hook

- Anchored to the cell wall and cell membrane by the basal body
- Cells with flagella:
    - Rotate flagella to run and tumble
    - Move toward or away from stimuli by sensing an increasing or decreasing concentration gradient (chemotaxis)

Axial Filaments (Endoflagella)

- Anchored at one end of a cell
- Covered by an outer sheath
- Rotation causes cell to move like a corkscrew through a cork

Fimbriae (Attachment Pilli)

- Shorter, straighter, thinner than flagella
- Not used for locomotion
- Allow bacteria to attach to surfaces
- Can be found at the poles of the cell, or covering the cell's entire surface

Sex Pili

- Longer than fimbriae
- Only one or two per cell
- Are used to transfer DNA from one bacterial cell to another

Cell Wall

Chemically and structurally complex, semi-rigid, provides structure, and protects the cell.

- Prevents osmotic lysis
- Contributes to the ability to cause disease in some species
    - Target for some antibiotics
- Made of peptidoglycan in bacteria
    - Polymer of a disaccharide
    - N-acetylglucosamine (NAG) & N-acetylmuramic acid (NAM)

Gram-positive Bacteria Cell Walls

- Contain many layers of peptidoglycan
    - Results in a thick and rigid structure
- Teichoic acids
    - May regulate movement of cations
    - Involved in cell growth
        - Prevent extensive wall breakdown and lysis

Gram-negative Cell Walls

- Contains only one to a few layers of peptidoglycan
- No teichoic acids in Gram-negative cell walls
- More susceptible to rupture than Gram-positive cells
- Outer membrane:
    - Composed of lipopolysaccharides, lipoproteins, and phospholipids
        - Lipopolysaccharide is composed of O polysaccharide (antigen)
        - Can be used to ID certain Gram-negative bacterial species
    - Protects the cell from harmful substances
    - Contains transport proteins called porins
    - Lipid A (endotoxin) can cause shock, fever, and even death if enough is released into the host's blood

Gram-negative Cell Wall:

Damage to Prokaryotic Cell Walls

Prokaryotic cell walls contain substances not normally found in animal cells so antibiotics that disrupt prokaryotic cell wall structures are often used. Example: Penicillin inhibits the formation of peptide bridges in peptidoglycan.

- Protoplast - Gram-positive cell with a destroyed cell wall
    - Still alive and functional
- Spheroplast – Gram-negative cell with no cell wall
- L forms - cells that swell into irregular shapes because they lack a cell wall
    - Can live, divide, and may return to a walled state
- Gram-negative bacteria are not as susceptible to penicillin
    - Due to their outer membrane and the small amount of peptidoglycan in their wall
    - Susceptible to antibiotics that can penetrate the outer membrane

Structures Internal to the Cell Wall

Plasma Membrane (Inner Membrane)

- Phospholipid bilayer
    - Basic framework of the plasma membrane
    - Bilayer arrangement occurs because the phospholipids are amphipathic molecules
        - Amphipathic molecules - contain polar and nonpolar portions
- Where photosynthesis, aerobic cellular respiration, and anaerobic cellular respiration reactions occur
- Selective permeability allows the passage of some molecules but not others across the plasma membrane
    - Large molecules cannot pass through
    - Ions pass through very slowly or not at all
    - Lipid soluble molecules and smaller molecules pass through easily
- Peripheral proteins
    - Enzymes and structural proteins
    - Some assist the cell in changing membrane shape
- Integral proteins and transmembrane proteins
    - Provide channels for movement of materials into and out of the cell
        - Transporters are used to pass material into or out of the cell

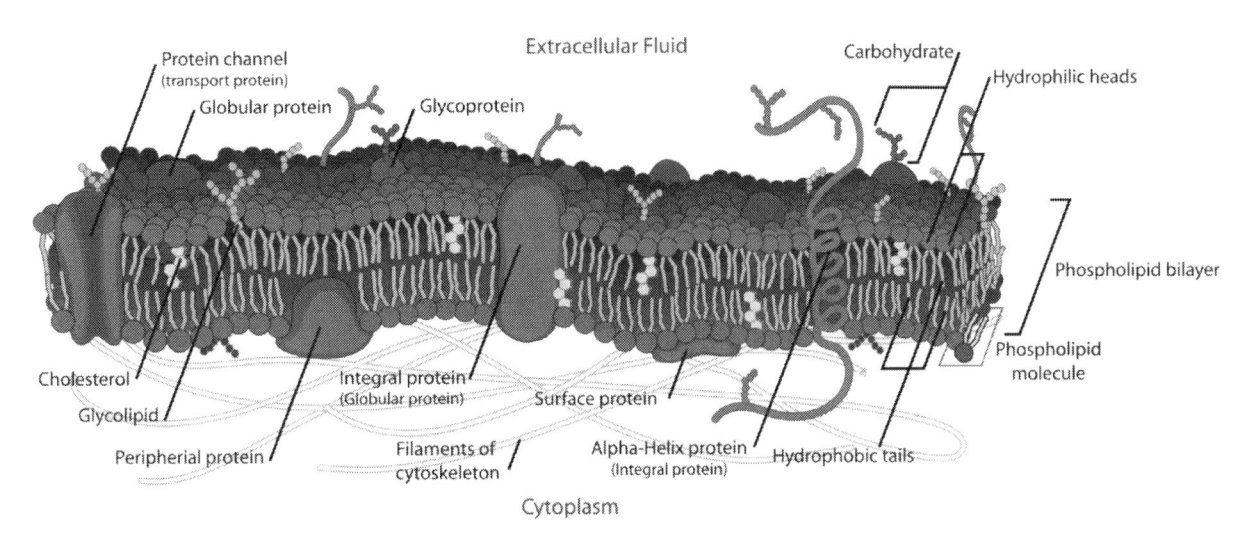

Cytoplasm

The cytoplasm is a substance inside the plasma membrane that is ~80% water. It contains proteins, enzymes, carbohydrates, lipids, inorganic ions, various compounds, a nuclear area, ribosomes, and inclusions.

Nuclear Area (Nucleoid)

There is a single circular chromosome made of DNA that has no histones or introns in bacteria. The chromosome is attached to the plasma membrane at a point along its length, where proteins synthesize and partition new DNA for division during binary fission.

Plasmids

Plasmids are small circular DNA molecules and are used in genetic engineering.

- Plasmids can be gained or lost without harming the cell
- Usually contain less than 100 genes
- Can be beneficial
    - May contain genes for antibiotic resistance, tolerance to toxic metals, production of toxins, or synthesis of enzymes
- Can be transferred from one bacterium to another

Ribosomes

Ribosomes are the site of protein synthesis and are composed of a large and small subunit, both made of protein and rRNA.

- Prokaryotic ribosomes are 70S ribosomes
    - Made of a small 30S subunit and a larger 50S subunit
- Eukaryotic ribosomes are 80S ribosomes
    - Made of a small 40S subunit and a larger 60S subunit
- Certain antibiotics target only prokaryotic ribosomal subunits without targeting eukaryotic ribosomal subunits

Inclusions

Inclusions are reserve deposits of nutrients that can be used in times of low resource availability.

Endospores

Endospores are dormant, tough, and non-reproductive structures produced by some Gram-positive bacterial cells that form when essential nutrients can no longer be obtained.

- Resistant to environmental hazards
- Sporulation (sporogenesis) - process of endospore formation within vegetative cells
    - Endospore is metabolically inert
        - Contains the chromosome, some RNA, ribosomes, enzymes, other molecules, and a very small amount of water
        - Can remain dormant for millions of years
- Germination - return to the vegetative state
    - Triggered by damage to the endospore coat

Eukaryotic Cell Structures

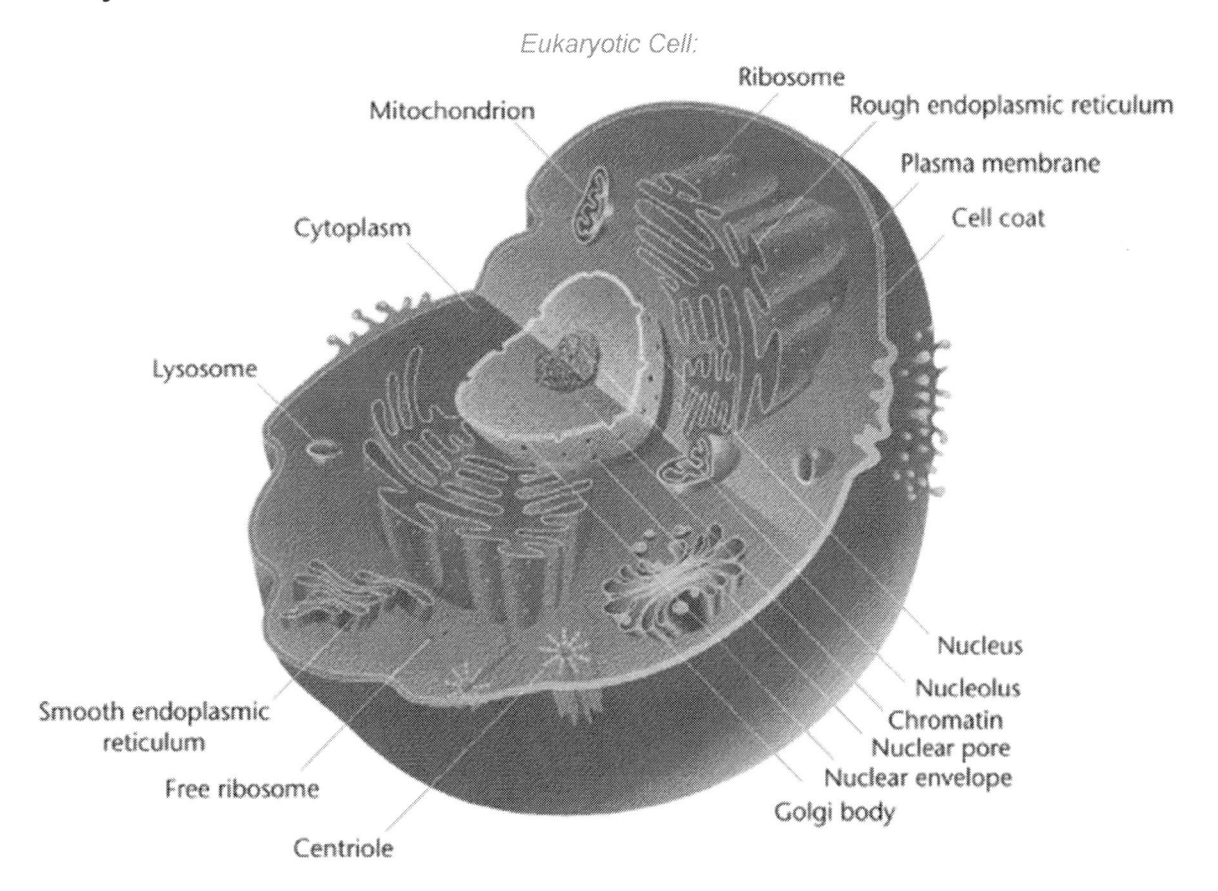

Eukaryotic Cell:

Labels: Mitochondrion, Ribosome, Rough endoplasmic reticulum, Plasma membrane, Cell coat, Cytoplasm, Lysosome, Nucleus, Nucleolus, Chromatin, Nuclear pore, Nuclear envelope, Golgi body, Smooth endoplasmic reticulum, Free ribosome, Centriole

Cell Wall

- Simpler in comparison to prokaryotes
    - No peptidoglycan in eukaryotes
- Cell walls are found in plants, algae, and fungi

Glycocalyx

Glycocalyx are sticky carbohydrates extending from an animal cell's plasma membrane. Helps cells recognize one another, adhere to one another in some tissues, and protects the cell from digestion by enzymes in the extracellular fluid.

Plasma Membrane

The plasma membrane is a flexible, sturdy barrier that surrounds and contains the cytoplasm of the cell. The membrane consists of proteins in a sea of phospholipids.

- Phospholipid bilayer
    - Same arrangement as the prokaryotic plasma membrane
- Membrane proteins are divided into integral and peripheral proteins
    - Integral proteins extend into or across the entire lipid bilayer
    - Peripheral proteins associate loosely with the polar heads of membrane lipids
        - Can be at the inner or outer surface of the membrane
    - Many membrane proteins are glycoproteins
        - Glycoproteins - proteins with carbohydrate groups attached to the ends that protrude into the extracellular fluid)
- Functions of membrane proteins:
    - Ion channels (pores) - Allow ions to cross the cell membrane
        - Most are selective - allow only a specific ion to pass
        - Some ion channels can open and close
    - Transporters - selectively move a polar substance from one side of the membrane to the other
    - Receptors - recognize and bind a specific molecule
        - Chemical binding to the receptor is called a ligand

- Enzymes – catalyze chemical reactions
- Cell-identity markers (often glycoproteins and glycolipids)
- Anchor proteins in the plasma membrane
- Has same active, passive, osmosis, vesicular transport capabilities as a prokaryotic plasma membrane
    - Vesicle is a small membranous sac
    - Two types of vesicular transport are endocytosis and exocytosis
    - Endocytosis
        - Materials move into a cell in a vesicle formed from the plasma membrane
        - Phagocytosis is the ingestion of solid particles
        - Pinocytosis is the ingestion of extracellular fluid
    - Exocytosis
        - Membrane-enclosed structures called secretory vesicles form inside the cell
        - They fuse with the plasma membrane and release their contents into the extracellular fluid

Cytoplasm

- Cytosol is the fluid portion of cytoplasm
- Mainly composed of water, salts, and proteins

Cytoskeleton

- Network of several kinds of protein filaments that extend throughout the cytoplasm
    - Provide the structural framework for the cell
- Consists of microfilaments, intermediate filaments, and microtubules
    - Most microfilaments are composed of actin and function in movement
        - Muscle contraction and cell division
    - Intermediate filaments are composed of several different proteins
        - Function in support and helping anchor organelles

- Microtubules are composed of a protein called tubulin and help determine cell shape
- Function in the intracellular transport of organelles and the migration of chromosome during cell division

Organelles

Organelles are specialized structures that have characteristic shapes and perform specific functions in eukaryotic cellular growth, maintenance, and reproduction.

Nucleus

- Most cells have a single nucleus
    - Some cells (red blood cells) have none
    - Others (human skeletal muscle fibers) have several in each cell
- Nuclear envelope containing channels called nuclear pores
    - Pores control the movement of substances between the nucleus and the cytoplasm

Ribosomes

- Sites of protein synthesis
- Membrane-bound ribosomes found on rough ER

Endoplasmic Reticulum

- Network of membranes extending from the nuclear membrane that form flattened sacs or tubules
- Rough ER has its outer surface covered with ribosomes
- Smooth ER extends from the rough ER to form a network of membrane tubules
    - Does not contain ribosomes

Golgi Complex

- Consists of four to six stacked, flattened membranous sacs (cisterns)
- Cis, medial, and trans cisternae each contain different enzymes
    - Allows them to modify, sort, and package proteins received from the rough ER for transport to different destinations

Lysosomes
- Lysosomes contain powerful digestive enzymes
    - Digest worn-out organelles (autophagy)
    - Digest their own cellular contents (autolysis)
    - Carry out extracellular digestion

Vacuoles
- Temporary storage for biological molecules and ions
- Bring food into cells
- Provide structural support
- Store metabolic wastes

Peroxisomes
- Similar in structure to lysosomes, but smaller
- Contain enzymes (oxidases) that use molecular oxygen to oxidize various organic substances
- Take part in normal metabolic reactions such as the oxidation of amino acids and fatty acids

Centrosomes
- Centrosomes are dense areas of cytoplasm containing the centrioles (paired cylinders arranged at right angles to one another)
    - Centrioles serve as centers for organizing microtubules and the mitotic spindle during mitosis

Mitochondria
- Cristae – folds of inner membrane
    - Highly variable in structure
- Outer membrane
    - Permeable to small molecules and ions

- Inner Membrane
    - Impermeable to most small molecules
    - Vital for forming and maintaining proton gradient
    - Carries ETS, ATP synthase, and translocase to move ADP in and ATP out
- Matrix
    - Contains oxidation enzymes
        - Except glycolytic enzymes
    - Also contains DNA, ribosomes, other enzymes, metabolic intermediates, small molecules, and ions
    - Inner membrane keeps it separate from cytosolic components
- Intermembrane space
    - Area between the inner and outer membranes

Chloroplasts (Only in Algae and Green Plants)

- Contain the pigment chlorophyll and enzymes necessary for photosynthesis

Endosymbiotic Theory

Large bacterial cells lost their cell walls and engulfed smaller bacteria and a symbiotic (mutualistic) relationship developed between them. The host cell supplied the nutrients, and the engulfed cell produced excess energy that the host could use. Evidence for the theory:

- Mitochondria and chloroplasts resemble bacteria in size and shape
- Divide on their own - independent of the host and the process is nearly identical to binary fission
- Contain their own DNA
    - Single circular chromosome
- Contain 70S ribosomes

Types of Transporters

Transporters may work through passive or active transport and are classified by the way in which they function.

- Uniporter - moves one substance at a time
- Cotransport – a single ATP-powered pump that actively transports one solute and indirectly drives the transport of other solutes against their concentration gradients
    - Symporter - Moves two substances in the same direction at the same time
    - Antiporter - Moves two substances in opposite directions at the same time

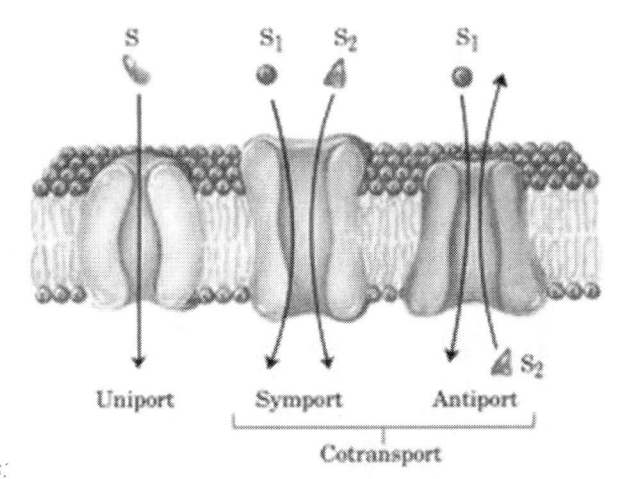

Types of Transporters:

Active vs. Passive Transport

- Active transport - energy is needed (nonspontaneous process) for a transport protein to pump a substance against its concentration gradient
- Passive transport - diffusion across a membrane, no energy needed (spontaneous)
    - Flow down concentration gradient
        - Concentration gradient – gradual difference of concentration over a distance in a particular direction
    - Passive diffusion (simple diffusion)
        - Diffusion – net movement of a substance down a concentration gradient
        - No pore or transport protein is needed
    - Facilitated diffusion
        - Transport proteins are used to pass material into or out of the cell
        - Transport proteins in facilitated diffusion do not use energy
            - Substance being transported travels down its concentration gradient

CHAPTER 5: MICROBIAL METABOLISM

Catabolism, Anabolism, and Metabolism

Metabolism is the sum of the chemical reactions in an organism. A metabolic pathway is a sequence of enzymatically catalyzed chemical reactions in a cell.

- Catabolism refers to energy-releasing processes
    - Catabolism provides the building blocks and energy for anabolism
- Anabolism refers to energy-using processes
- ATP participates in coupling anabolic and catabolic reactions

Enzymes

Metabolic pathways are regulated and determined by enzymes.

- Chemical reactions can occur when atoms, ions, and molecules collide
- Activation energy is needed to disrupt electronic configurations
    - Enzymes reduce the activation energy of a chemical reaction
- Reaction rate is the frequency of collisions with enough energy to bring about a reaction
    - Reaction rate can be increased by increasing enzyme concentration, temperature or pressure
- Enzymes are biological catalysts
    - Each enzyme is specific to a single chemical reaction
        - Enzyme is not used up or altered in that reaction
        - Specificity determined by the 3-D shape of the enzyme
    - An enzyme only acts on a specific substrate
- Names of enzymes normally end in –ase

Classes of Enzymes

- Oxidoreductases – involved in oxidation-reduction reactions
- Transferases – involved in transfer of functional groups

- Hydrolases – involved in hydrolysis reactions
- Lyases – perform group elimination to form double bonds
- Ligases – involved in bond formation coupled with ATP hydrolysis
- Isomerases – involved in isomerization reactions

Enzyme Components

- Apoenzyme – inactive enzyme that requires activation before functioning
- Cofactor - non-protein component that helps catalyze a reaction by forming a bridge between the enzyme and substrate
- Holoenzyme: Apoenzyme + cofactor OR coenzyme
    - Important Coenzymes:
        - NAD^+, $NADP^+$, FMN, FAD, Coenzyme A

Mechanism of Enzymatic Action

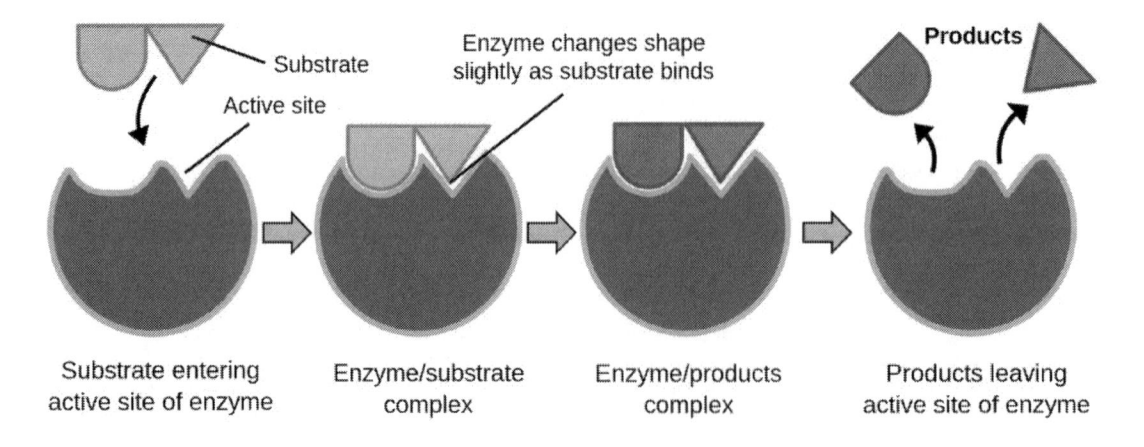

- Substrate contact a specific part of the enzyme called the active site
    - Enzyme-substrate complex forms
- Substrate is transformed by one of the following:
    - Rearrangement of existing atoms
    - Breakdown of the substrate molecule (hydrolysis)
    - By combining with another substrate molecule (dehydration synthesis)

- Transformed substrate is released from the enzyme
  - Substrate no longer fit in the active site
  - Enzyme is free to catalyze the same reaction again

Factors Influencing Enzyme Activity

- Rate of enzyme synthesis
- Temperature
- pH
- Substrate concentration
- Denatured enzymes

Competitive Inhibition

- Compound resembles substrate or transition state
  - Those resembling the transition state are usually more potent
- Occupies the active site
- Adding more substrate can out-compete inhibitor
- No change in catalysis
  - Once a substrate binds you get product at the same rate
- Competitive inhibitor does not undergo a reaction
  - Some competitive inhibitors bind reversibly while others bind irreversibly

Noncompetitive Inhibition

Noncompetitive inhibitors don't bind at the active site; instead, they bind with another part of the enzyme called the allosteric site. This binding causes the active site to change shape, making it non-functional. This is called allosteric inhibition. This can be reversible or non-reversible.

Feedback Inhibition

In many metabolic pathways where multiple enzymes catalyze a series of reactions in sequence, the end-product can allosterically inhibit the activity of an enzyme earlier in the pathway.

- This inhibition is reversible
- Stops the cell from wasting chemical resources by making more of a substance than it needs
- As the end-product gets used up, the enzyme's allosteric site will more often remain unbound, and the pathway can resume activity

Ribozymes

Ribozymes are RNA molecules capable of acting as enzymes. They cut and splice RNA. They function like enzymes in that they have active sites that bind to substrates and are not used up in a chemical reaction.

Energy Production

Oxidation-Reduction Reactions

- Oxidation = loss of electrons
- Reduction = gain of electrons
- Trick for remembering - OIL RIG
    - **OIL** - **O**xidation **I**s **L**osing electrons
    - **RIG** - **R**eduction **I**s **G**aining electrons
- Redox reactions are oxidation reactions paired with reduction reactions

Generation of ATP

- Addition of PO_4^{-3} (inorganic phosphate = P_i) to a compound is called a phosphorylation
- ATP is generated by the phosphorylation of ADP
- Substrate-level phosphorylation
    - Transfer of a high-energy P_i from a phosphorylated compound directly to ADP

- Oxidative phosphorylation
    - Energy released from the sequential transfer of electrons of one compound to another is used to generate ATP by chemiosmosis via oxidative phosphorylation in the electron transport chain (ETS)
    - Molecular oxygen (O_2) is the final electron acceptor in aerobic oxidative phosphorylation (aerobic cellular respiration)
- Photophosphorylation
    - Light causes chlorophyll to give up electrons
    - Energy released from the transfer of electrons of chlorophyll through a system of carrier molecules via an electron transport chain is used to generate ATP and NADPH
    - ATP and NADPH are then used to synthesize organic molecules from CO_2 and H_2O (Carbon fixation)

Carbohydrate Catabolism

Carbohydrate catabolism is the breakdown of carbohydrates to release energy. Most organisms oxidize carbohydrates as their primary source of cellular energy. Glucose is the most common energy source, and cellular respiration and fermentation are the two general processes microorganisms use to produce energy from glucose.

- Cellular respiration can be aerobic or anaerobic
- Cellular respiration involves:
    - Glycolysis
    - Krebs cycle
    - Electron transport chain
- Fermentation typically begins with glycolysis
    - However, it does not include the Krebs cycle or ETC
        - Therefore the ATP yield is much lower
    - Fermentation is considered a separate process from cellular respiration

Glycolysis

In glycolysis glucose is oxidized to pyruvic acid, producing ATP and NADH + H⁺. Glycolysis is broken down into 2 stages.

- Energy Investment Phase
    - 2 ATPs are used
    - 5 steps that result in glucose being split to form a pair of glyceraldehyde-3-phosphate molecules
- Energy Payoff Stage:
    - 5 steps that result in the pair of glyceraldehyde-3-phosphate molecules becoming oxidized to form a pair of pyruvic acid molecules
    - 4 ATP molecules are produced by substrate-level phosphorylation
    - 2 NADH + 2 H⁺
- Net equation of glycolysis
    - Glucose + 2 NAD⁺ + 2 P_i → 2 pyruvate + 2 NADH + 2 ATP

Cellular Respiration

Cellular respiration is the process in which cells break up sugars into a form that the cell can use as energy. There are two types of cellular respiration: aerobic and anaerobic.

- Oxidation of molecules liberates electrons to be used in the electron transport chain
- ATP is generated by oxidative phosphorylation in the electron transport chain and by substrate-level phosphorylation in the Krebs Cycle
- Intermediate Step:
    - Pyruvic acid (from glycolysis) which is a 3-carbon compound, is oxidized and decarboxylated giving off a molecule of CO_2
        - Results in the formation of a 2-carbon compound called an acetyl group
        - A molecule of NAD⁺ is reduced to NADH + H⁺
    - Acetyl group attaches to coenzyme A to form acetyl coenzyme A (acetyl CoA)
    - Because glycolysis ended in the formation of two pyruvic acid molecules from a single glucose molecule, these steps happen twice

Aerobic Cellular Respiration

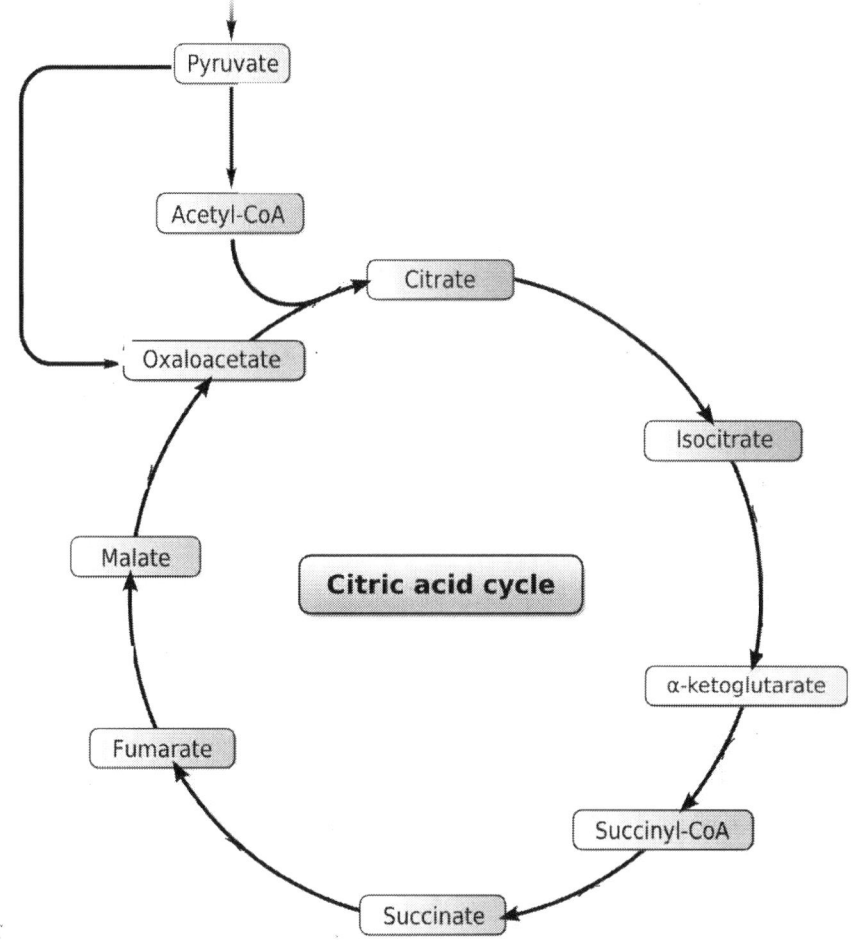

- Krebs Cycle

Krebs Cycle:

- Acetyl CoA enters the Krebs cycle, CoA detaches, and the acetyl group combines with oxaloacetic acid (a 4-carbon molecule) to form citric acid (a 6-carbon molecule)

- Through a series of reactions, the Krebs cycle produces:
    - 3 NADH + 3H$^+$
    - 1 FADH$_2$
    - 1 ATP via substrate level phosphorylation
    - 2 CO$_2$

- This process occurs twice, because 1 glucose → 2 pyruvic acid → 2 acetyl CoA molecules

- All remaining carbon atoms of the original glucose molecule are individually cleaved, forming CO_2
- Krebs cycle ends with the formation of oxaloacetic acid
    - Which is then free to bind another acetyl group and start the sequence over again
- 3 NADH + 3H⁺ and 1 FADH₂ produced are used in the electron transport chain

Electron Transport Chain (ETS)

The ETC is a series of carrier molecules that are oxidized and reduced as electrons are passed down the chain.

- 3 classes of carrier molecules:
    - Flavoproteins - contain flavin, a coenzyme derived from riboflavin (vitamin B2), includes FMN (Flavin mononucleotide)
    - Cytochromes - proteins with a heme group
    - Ubiquinones - aka coenzyme Q, which are small non-protein carriers
- NADH is oxidized to NAD⁺ when it releases 2 electrons and a proton to FMN, the first carrier in the ETC
- Another proton follows from the surrounding aqueous medium
- Protons enter the space between the inner and outer mitochondrial membranes (intermembranous space)
    - Meanwhile the electrons get passed from one carrier molecule to the next in a series of redox reactions
    - As electrons are passed through the chain, there is a stepwise release of energy, which is used to drive the chemiosmotic generation of ATP
- Last cytochrome in the ETC, cyt a_3, passes its electrons to molecular oxygen (O_2)
    - Adding electrons to oxygen causes it to become negatively charged
        - It picks up protons (H+) from the surrounding medium to form H_2O
    - Molecular oxygen (O_2) is the final electron acceptor in aerobic cellular respiration

- As the electrons move down the ETC more protons follow and are released into the intermembranous space of the mitochondrion
    - Other protons are pumped into this space from the mitochondrial matrix
    - This generates both a concentration gradient and an electrical gradient across the inner mitochondrial membrane
        - This is called the proton motive force
    - Protons can diffuse from the intermembranous space across the inner mitochondrial membrane and into the mitochondrial matrix through special protein channels that contain ATP synthase
    - Movement of protons is what drives the formation of ATP from ADP and Pi
- Single molecule of NADH produces 3 ATP during oxidative phosphorylation due to chemiosmosis
- Single molecule of FADH$_2$ produces 2 ATP during oxidative phosphorylation due to chemiosmosis
    - Difference in ATP production is because the electrons carried by FADH$_2$ enter the ETC at a lower point than the electrons delivered by NADH
- ETC is found on the inner mitochondrial membrane of mitochondria in eukaryotes and on the plasma membrane of prokaryotes
- ETC regenerates NAD$^+$ and FAD, which can be used again in glycolysis and the Krebs cycle
- $C_6H_{12}O_6 + 6\ O_2 + 38\ ADP + 38\ P_i \rightarrow 6\ CO_2 + 6\ H_2O + 38\ ATP$
    - 36-38 ATPs are produced in eukaryotes
    - 38 ATPs are produced in prokaryotes

Anaerobic Cellular Respiration

Final electron acceptor in the electron transport chain is not O_2. Instead, inorganic substances other than oxygen, such as nitrate ions (NO_3^-) serve as the final electron acceptor. Rarely, an organic molecule serves as the final electron acceptor.

- Glycolysis proceeds exactly as mentioned earlier
- Anaerobic cellular respiration yields less energy than aerobic cellular respiration
    - Only part of the Krebs cycle operates under anaerobic conditions
    - Also not all the carriers in the ETC participate in anaerobic cellular respiration

- ATP yield is less than 38 but more than 2

Fermentation

Glycolysis proceeds as previously discussed, producing 2 pyruvic acid molecules, 2 net ATP, and 2 NADH + 2H⁺.

- Pyruvic acid is then converted to another organic molecule, and NAD⁺ is regenerated so it can participate in another round of glycolysis
    - Alternatively, energy can be derived from other organic molecules such as amino acids, organic acids, purines, and pyrimidines
- Does not require oxygen
- Krebs cycle and ETC do not occur
- Uses an organic molecule as the final electron acceptor
- Produces only small amounts of ATP, which are generated during glycolysis
- Lactic acid fermentation
    - Glycolysis produces 2 ATP and 2 pyruvic acid molecules
        - They are reduced by 2 NADH to produce 2 lactic acid molecules and 2 NAD⁺
- Alcohol fermentation
    - Glycolysis produces 2 ATP and 2 pyruvic acid molecules
        - They are converted to 2 CO_2 molecules and 2 acetaldehyde molecules
    - The 2 acetaldehyde molecules are reduced by 2 NADH to produce 2 NAD⁺ and 2 ethanol molecules

Lipid and Protein Catabolism

Organisms can also use lipids and proteins as sources of energy, not just glucose.

Lipid Catabolism

- Microbes produce lipases (extracellular enzymes) that break down tri-, di-, and monoglycerides into glycerol and fatty acid
    - Glycerol enters the glycolysis pathway
    - Fatty acids undergo beta oxidation
        - 2-carbon fragments are cleaved from the fatty acids and bond with CoA to form acetyl-CoA, which then enters the Krebs cycle

Protein Catabolism

Microbes produce proteases and peptidases (extracellular enzymes) that break proteins down into amino acids, which can then enter the cell.

- Before amino acids can be catabolized, they must be broken into molecules that can enter the Krebs cycle
- Deamination is the process where the amino group is removed from the carbon backbone of the amino acid and is converted to NH_4^+ (an ammonium ion)
    - Can then be excreted as a waste product by the cell
- Remaining organic acid can enter the Krebs cycle
- Other conversions involve decarboxylations (removal of –COO) and dehydrogenations

Photosynthesis

Photosynthesis is the process in which light energy from the sun is converted into chemical energy. The chemical energy is then used to convert CO_2 from the atmosphere into more reduced carbon compounds, primarily sugars. This is also known as carbon fixation.

- Oxygenic photosynthesis equation
    - $6 CO_2 + 12 H_2O + \text{Light (photon)} \rightarrow C_6H_{12}O_6 + 6 O_2 + 6 H_2O$
- Water is the electron donor

Light-dependent (Light) Reactions: Photophosphorylation

- Cyclic photophosphorylation (a relatively uncommon photosynthesis pathway that does not generate oxygen gas)
    - Light excites a pair of electrons from chlorophyll
        - Chlorophyll is found in the membranous thylakoids of chloroplasts
    - The excited electrons (existing in a higher energy state) enter an electron transport chain
    - As the electrons move down the chain, protons are pumped across the membrane (proton motive force)
        - ADP and P_i are converted to ATP by chemiosmosis
        - Electrons return to the chlorophyll molecule
- Non-cyclic photophosphorylation (more common pathway)
    - Light excites a pair of electrons from chlorophyll
    - Excited electrons enter an electron transport chain
        - Electrons will not return to chlorophyll but will become incorporated into NADPH
        - Electrons lost from chlorophyll will be replaced by electrons from some reducing substance
            - H_2O in oxygenic photosynthesis
            - H_2, or H_2S in anoxygenic photosynthesis
    - As the electrons move down the chain, protons are pumped across the membrane
        - ADP and P_i are converted to ATP by chemiosmosis.
    - Electrons are transferred to the electron carrier $NADP^+$ which is reduced to NADPH

Light-independent (Dark) Reactions: Calvin-Benson Cycle

- Electrons bound up in NADPH and ATP are used to reduce CO_2 to form sugar molecules
- To form one molecule of glucose, the cycle must run six times
    - Requires a total investment of 6 CO_2, 18 ATP, and 12 NADPH

- 3 runs of the cycle produce a single glyceraldehyde 3-phosphate
- A pair of glyceraldehyde-3-phosphate molecules are required to form a single molecule of glucose

Metabolic Diversity Among Organisms

- Phototrophs - generate energy from light
- Chemotrophs - can generate energy from organic or inorganic chemicals
- Autotrophs - use CO_2 as their principal carbon source
- Heterotrophs - require an organic carbon source
- Photoautotrophs Types
    - Oxygenic photoautotrophs
        - Energy is used in the Calvin-Benson cycle to fix CO_2
        - H_2O is the electron donor used to reduce CO_2
    - Anoxygenic photoautotrophs
        - $CO_2 + 2\ H_2S + Photon \rightarrow CH_2O + 2\ S + H_2O$
        - Sulfur containing compounds are used to reduce CO_2
        - These organisms cannot carry on photosynthesis when oxygen is present and cannot use H_2O to reduce CO_2
- Photoheterotrophs
    - Use light as the energy source and organics such as alcohols, fatty acids, other organic acids, and carbohydrates as the carbon source
- Chemoautotrophs
    - Use electrons in reduced inorganic compounds as the energy source and CO_2 as the principal carbon source
- Chemoheterotrophs
    - Energy source and carbon source are usually the same organic compound
        - Electrons from the hydrogen atoms of the organic compound are the specific energy source
    - Many bacteria and fungi that can use a wide variety of organics as the carbon and energy source fall into this group

Metabolic Pathways of Energy Use

Only a few steps are needed to convert polymers to monomers but many steps are required to catabolize monomers or anabolize them.

- A metabolic pathway is series of chemical reactions needed to break down or build up monomers
    - Each step generally involves a different enzyme
    - Always regulated in some way
- Purpose of many steps
    - Convert large energy source into multiple smaller sources or energy carriers
        - Less chance that energy is wasted that way
        - Smaller "packets" means cells don't spend more than what they need
        - Multiple steps means multiple intermediates or metabolites

CHAPTER 6: MICROBIAL GROWTH

Requirements of Growth

Microbial growth is the increase in number of cells, not cell size.

Physical Requirements

- Temperature
    - Psychrophiles – grow in temperature less than 10 °C
        - Found deep in the ocean or in polar regions
    - Psychrotrophs - also cold loving microbes
        - Grow at 0°C but optimally at 20-30°C
    - Mesophiles - prefer moderate temperature
        - Grow optimally between 25-40 °C
        - Most common types of microbes
    - Thermophiles – prefer hotter environments
        - Many grow optimally between 50-60 °C.
    - Hyperthermophiles – grow optimally at more than 80 °C
        - Found in geothermal hot springs and deep ocean hydrothermal vents
- pH
    - Acidophiles – pH less than 5.5
    - Neutrophiles – pH 5.5 to 8.5
    - Alkalophiles – pH greater than 8.5

- Osmotic Pressure
  - When a microbe is suspended in a hypertonic solution, water moves out of the cell via osmosis
    - Causes plasmolysis - shrinkage of the cell's cytoplasm
      - Can permanently damage the plasma membrane
    - Growth is inhibited as the plasma membrane pulls away from the cell wall
  - Extreme and obligate halophiles require high salt concentrations in order to grow
  - Facultative halophiles do not require high salt concentrations
    - However, they can grow in salt concentrations up to 2%, which inhibits the growth of many microbes
  - When a microbe is suspended in a hypotonic solution, water enters the cell via osmosis
    - If the microbe has a relatively weak cell wall, it may lyse (osmotic lysis)

Chemical Requirements

- Carbon
  - Used to make structural organic molecules and energy rich molecules
  - Heterotrophs use organic carbon sources
  - Autotrophs use CO_2
- Nitrogen
  - Found in amino acids, proteins, DNA, RNA, ATP
  - Most bacteria decompose proteins
  - Some bacteria use ammonium ions (NH_4^+) or nitrate ions (NO_3^-) as nitrogen sources
  - Few bacteria use atmospheric N_2 in nitrogen fixation

- Sulfur
    - Found in the R group of some amino acids and the vitamins thiamine and biotin
    - Also necessary for synthesis of DNA & RNA
    - Some bacteria use sulfate ions (SO_4^{2-}) or H_2S
- Phosphorus
    - Found in DNA, RNA, ATP, and phospholipids
    - Phosphate ion (PO_4^{3-}) is a common source of phosphorus
- Trace Elements
    - Inorganic elements required in small amounts
    - Usually needed to serve as enzyme cofactors
- Oxygen
    - Obligate aerobes -- require oxygen to live
        - Use molecular oxygen as the final electron acceptor in the ETC of the aerobic pathway
    - Microaerophiles - are aerobic but grow only when oxygen concentrations are lower than that found in air
    - Aerotolerant anaerobes – aren't harmed and don't use oxygen to grow
    - Facultative anaerobes – can use oxygen to grow but can also grow in its absence
    - Obligate anaerobes – cannot grow in the presence of oxygen

- Toxic Forms of Oxygen
  - Singlet oxygen: $^1O_2^*$ - oxygen at a high energy state
    - Extremely reactive
  - Superoxide free radicals: $O_2\cdot^-$
    - Has an unpaired electron
    - Highly unstable
    - Formed during normal aerobic cellular respiration when oxygen is used as the final electron acceptor
    - Steal electrons from neighboring molecules
      - Turns them into radicals
    - Neutralized by superoxide dismutase to produce H_2O_2.
  - Peroxide anion: O_2^{2-}
    - Two peroxide anions combine with 2 protons to form O_2 and H_2O_2
    - H_2O_2 is then neutralized by catalase to form water and O_2
  - Hydroxyl radical: $OH\cdot$
    - Produced by ionizing radiation and when H_2O_2 reacts with certain metal ions
- Organic Growth Factors
  - Organic compounds obtained from the environment that the organism can't synthesize itself
    - Vitamins, certain amino and fatty acids, purines, pyrimidines

Biofilms

Biofilms are complex communities of microorganisms living together. They can communicate with one-another to coordinate activities, similar to the way cells in a multicellular organism communicate.

Culture Media

- Culture medium - nutrients prepared for microbial growth
- Sterile - no living microbes
- Inoculum - introduction of microbes into a medium
- Culture - microbes growing in/on culture medium
- Agar - complex polysaccharide used as solidifying agent for culture media
    - Generally not metabolized by microbes
- Defined media - exact chemical composition is known
    - Normally used for the growth of autotrophic bacteria (use CO_2 as the carbon source)
- Complex media - extracts and digests of yeasts, meat, or plants that provide vitamins and organic growth factors
    - Used to grow heterotrophic bacteria and fungi (which require an organic carbon source)
    - Liquid form is called a nutrient broth
    - Solid form is called a nutrient agar
- Selective media - halts or inhibits the growth of a particular group of microorganism while not affecting the growth of another group of microorganism
- Differential media - provides a visual difference between groups of organisms that is not related to how well the organisms grow on the medium

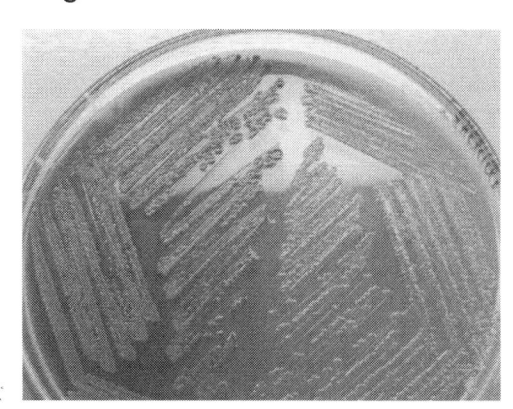

MacConkey Agar (Differential Media Example):

- Eosin Methylene Blue (EMB) - selective against Gram-positive bacteria, and differential on the basis of lactose fermentation

- Enrichment media - Encourages growth of desired microbe but not others, so it is also a selective medium

Obtaining Pure Cultures

A pure culture contains only one species or strain of microorganism. A colony is a population of cells arising from a single cell or spore or from a group of attached cells. A colony is often called a colony-forming unit (CFU).

Streak Plate Method

- Works well when the organism to be isolated is present in large numbers relative to the total microbial population
    - Otherwise, the organism must be grown in an enrichment medium first

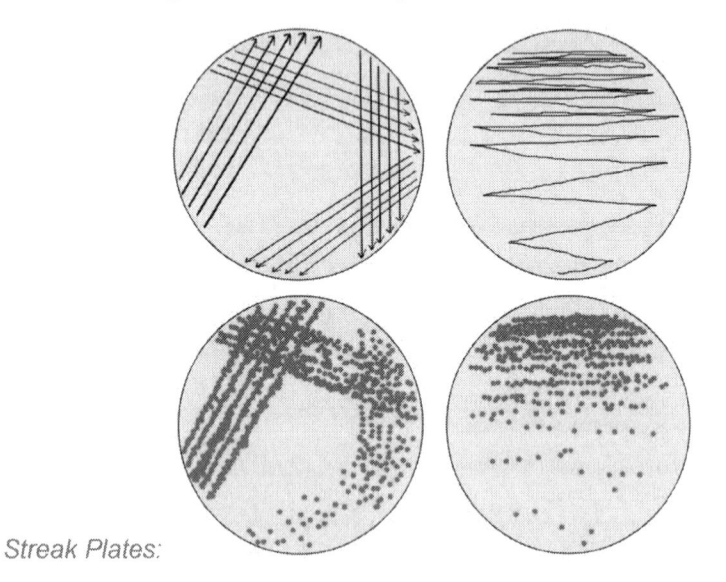

Streak Plates:

Preserving Bacterial Cultures

- Short-term storage is accomplished by refrigeration
- Long term storage:
    - Deep-freezing
        - Pure culture in a suspending liquid which is then quick frozen at very low temperature
    - Lyophilization (freeze-drying)
        - Culture is frozen and dehydrated in a vacuum (sublimation)

Growth of Bacterial Cultures (Phases of Growth)

- Lag Phase – cells adjust to the new environment and change gene expression
- Log (Exponential Phase) – maximum growth
    - Used for generation time
        - Time for population to double
    - Balanced growth
    - Reproducible
- Stationary Phase – nutrients decreasing, growth rate approaching the death rate
- Death phase – low nutrients, high death rate

Direct Measurements of Microbial Growth

Plate Counts

- Perform serial dilutions of a sample
- Inoculate Petri plates from serial dilutions
- After incubation, count colonies on plates that have 25-250 colonies (CFUs)
    - Count is used to estimate the number of bacteria in the original sample

Filtration

- Used when bacterial numbers are low within a liquid or gaseous medium
- Filter the liquid, then transfer the filter to a Petri dish containing a pad saturated in liquid nutrient medium
    - Colonies will grow on the filter's surface
    - Often used to detect bacteria in water bodies/supplies

Direct Microscopic Count

- Measured volume of liquid is placed on a defined area of a microscope slide and the bacteria present are counted
- Cons
    - Not good for counting motile bacteria
    - Dead bacterial cells are as likely to be counted with live ones
    - High concentration of cells is required

- Pros
    - No incubation time is required to make a count

Estimating Microbial Growth By Indirect Methods

- Turbidity – cloudiness or haziness of a fluid caused by large numbers of individual particles (e.g. cells) that is generally visible with the naked eye
- Metabolic activity
    - Assumes that a bacterial metabolic product accumulates in direct proportion to the number of bacteria present
- Dry weight
    - Used for filamentous bacteria and molds
    - Organism is filtered from the growth medium/sample, then dehydrated and weighed

Chapter 7: Controlling Microbial Growth

Terminology

- Sterilization - removal or destruction of all microbial life
    - Commonly performed by heating
- Disinfection - destruction of vegetative pathogens that do not form endospores
    - Accomplished by using (disinfectants), UV radiation, boiling water, etc.
- Antisepsis - removal of pathogens from living tissue
    - Antiseptic – compounds that are applied to living tissue to remove pathogens
- Degerming - mechanical removal of microbes from a limited area
    - Generally does not kill microorganisms
- Sanitization – cleaning something to make it free of bacteria
- Biocide/Germicide – kills microbes, but not usually endospores
- Fungicides - kill fungi
- Virucides – destroys or deactivates viruses
- Bacteriostasis - inhibits the growth of microbes without necessarily killing them
- Sepsis - microbial contamination
- Asepsis - absence of significant contamination
    - Aseptic surgery techniques prevent microbial contamination of wounds

Rate of Microbial Death

Bacterial populations die at a logarithmic rate when heated or treated with antimicrobial chemicals.

- Effectiveness of antimicrobial treatment depends on:
    - Number of microbes present at the beginning of the treatment
    - Environment (temperature, ph, etc.)
    - Time of exposure
    - Microbial characteristics

Actions of Microbial Control Agents

- Alteration of membrane permeability/cell wall integrity
- Damage to intracellular proteins
    - Denature proteins and enzymes by targeting hydrogen bonds or disulfide bridges
- Damage to nucleic acids

Physical Methods of Microbial Control

Heat

- Kills microorganisms by denaturing proteins and enzymes
- Thermal death point (TDP) - minimum temperature at which all cells in a liquid suspension are killed in 10 min
- Thermal death time (TDT) - Minimum time required to kill all cells in a liquid at a given temperature
- Decimal reduction time (DRT = D value) - Minutes to kill 90% of a population at a given temperature
- Moist heat denatures proteins and enzymes

- Autoclave - a strong heated container used for processes requiring high pressure and temperature
    - Increasing pressure raises the boiling point of water
        - Allows for much higher temperatures to be achieved
    - Steam at a pressure of about 15 psi (121°C) will kill all organisms and their endospores in about 15 minutes
        - Accomplished by denaturing proteins and enzymes
        - Sterilization of a surface requires that the steam directly contact it
- Dry heat sterilization kills by oxidation
    - Flaming, incineration, and hot-air sterilization

Filtration

- Removes microbes from liquids and gasses
- Pores in the filter allow the passage of the medium, but not the microbe
- Useful for sterilizing liquids that would be altered by heating

Cooling

- Refrigeration (0-7°C)
    - Slows the metabolic rate of most microbes so that they cannot reproduce or synthesize toxins
- Slow freezing
    - Allows ice crystals to form that can kill some species of bacteria

Pressure

- High pressure denatures proteins and alters carbohydrate structure, killing vegetative bacterial cells
- Endospores are relatively resistant to high pressure

Desiccation (State of Extreme Dryness)

- Prevents metabolism
- Cells can still remain viable for years
- Resistance to desiccation varies by species
  - Osmotic pressure causes plasmolysis

Radiation

- Ionizing radiation (X rays, gamma rays, electron beams) destroys DNA
  - Ionizes water
    - Produces hydroxyl radicals that react with organic molecules (especially DNA)
- Nonionizing radiation (UV) damages DNA
  - Causes bonds to form between adjacent pyrimidine bases, usually thymines, in DNA chains
  - Thymine dimers inhibit correct replication of DNA during cellular reproduction

Chemical Methods of Microbial Control

Types of Disinfectants

- Phenol and Phenolics
  - Phenol and phenolics damage the lipids of the plasma membrane
    - Cellular contents may leak
    - Also denature proteins and enzymes
  - Stable, and persist for long periods after application
- Bisphenols
  - Work by disrupting plasma membranes
  - Used in disinfectant hand soaps and skin lotions
- Biguanides
  - Disrupts plasma membranes
  - Used to disinfect skin

- Halogens
    - Iodine (I$_2$) inhibits protein function and is a strong oxidizing agent
    - Chlorine is the strong oxidizer in hypochlorous acid = HOCl (bleach) that alters various cellular components and affects pH
    - Iodine and chlorine are effective antimicrobial agents
        - Either alone or as components of other compounds
- Alcohols
    - Kill bacteria and fungi
        - But not endospores or non-enveloped viruses
    - Act by denaturing proteins and enzymes, and by dissolving lipids (plasma membrane, viral envelope)
- Heavy metals (i.e. Ag, Hg, Cu, Zn)
    - Can be biocidal or antiseptic
    - Denature proteins and enzymes by combining with sulfhydryl groups
- Surface-active agents or surfactants
    - Soaps
        - Emulsifies oils are used to degerm surfaces (like the skin)
    - Acid-anionic detergents
        - Thought to involve enzyme inactivation or disruption
    - Quaternary ammonium compounds (Quats)
        - Disrupt plasma membranes and denatures enzymes and proteins
- Chemical food preservatives
    - Added to food to retard spoilage
        - Inhibits metabolism (mostly of molds)
    - Organic acids or salts of organic acids that the body can metabolize
- Antibiotics
    - Injected or ingested to treat disease

- Aldehydes
    - Inactivate proteins and enzymes by cross-linking with functional groups
- Gaseous chemosterilizers
    - Denature proteins and enzymes
    - An object is placed in a chamber and the air is replaced with a gas
    - Good for sterilizing objects that would be damaged by heat or liquids
- Plasmas
    - An electromagnetic field creates radicals
    - Used to sterilize surgical instruments and tubing
- Supercritical fluids
    - Compressed CO_2
    - Has liquid and gaseous properties
    - Used on foods and donor tissues (bones, ligaments, tendons)
- Peroxygens (Oxidizing agents)
    - Oxidize cellular molecules

Viruses and Prions

Viruses

Enveloped viruses have an outer lipid envelope.

- Antimicrobials that are lipid soluble are more likely to be effective against enveloped viruses
- Non-enveloped viruses, which have only a protein coat, are more resistant; fewer

Prions

Prions are protein particles that are the agent of infection in a variety of neurodegenerative diseases. Prions are the only known infectious agents that do not contain DNA or RNA.

- Inactivation of prions
    - Protease enzymesm, phenol, incineration

CHAPTER 8: MICROBIAL GENETICS

Structure and Function of the Genetic Material

Genetics is the study of what genes are, how they carry information, how information is expressed, how genes/DNA are replicated, and how they are passed from one generation to the next (or passed between organisms).

- Chromosome - an entire DNA molecule

- Gene - segment of DNA (or RNA in some viruses) that encodes a functional product (usually a protein)

- Genome - all of the genetic material in a cell/organism

- Genomics - molecular study of genomes

- Genotype - entire DNA complement of an organism/cell (combination of alleles)

- Phenotype - physical expression of the genes (proteins/enzymes)

DNA

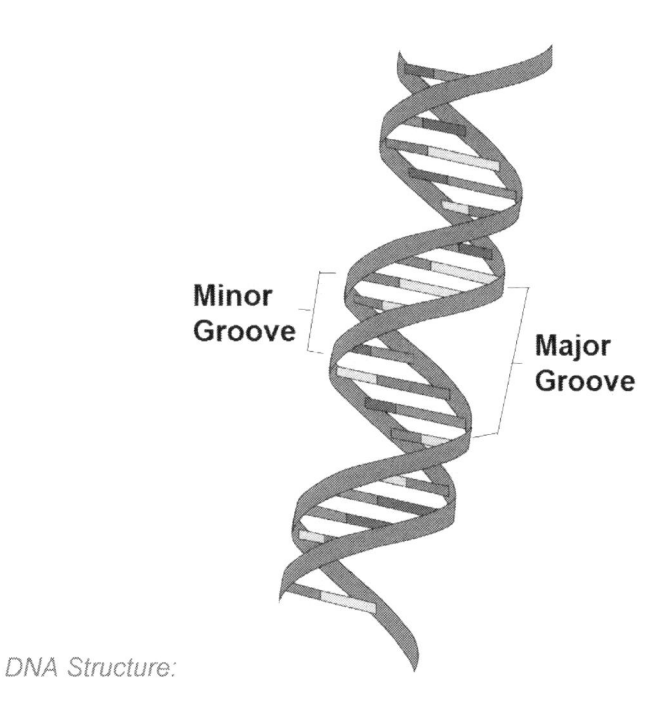

DNA Structure:

Structure of DNA

- Polymer of nucleotides
    - Adenine (A), thymine (T), cytosine (C), and guanine (G)

- Double helix
- DNA "backbone" is in a repeated pattern: deoxyribose, phosphate, deoxyribose, phosphate...
- Strands held together by hydrogen bonds between nitrogenous base pairs
    - A pairs with T
        - Two hydrogen bonds between them
    - C pairs with G
        - Three hydrogen bonds between them
- Two strands run antiparallel to one another
    - 5' to 3' pairs with 3' to 5'

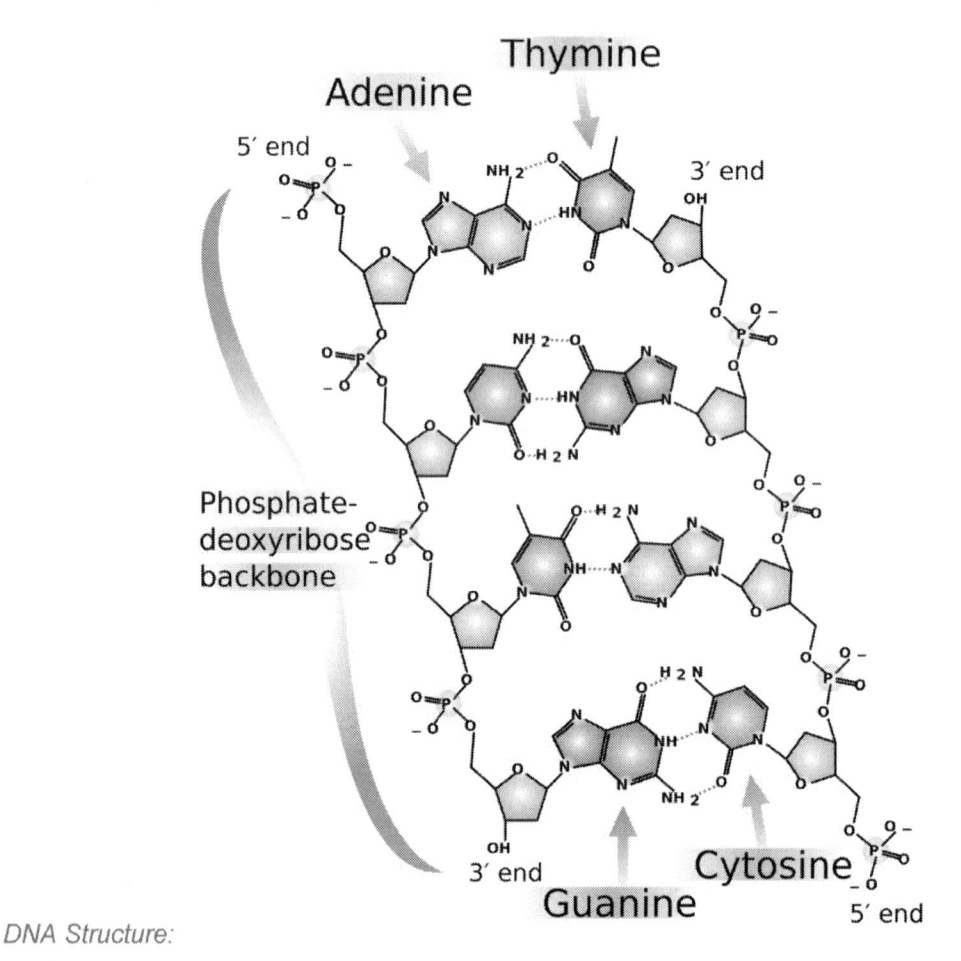

DNA Structure:

- Base pairs are perpendicular to the backbone
    - Base pairs stack closer together
    - Movement of bases together tilts the backbone by 30°
        - Creates an uneven twist

DNA Replication

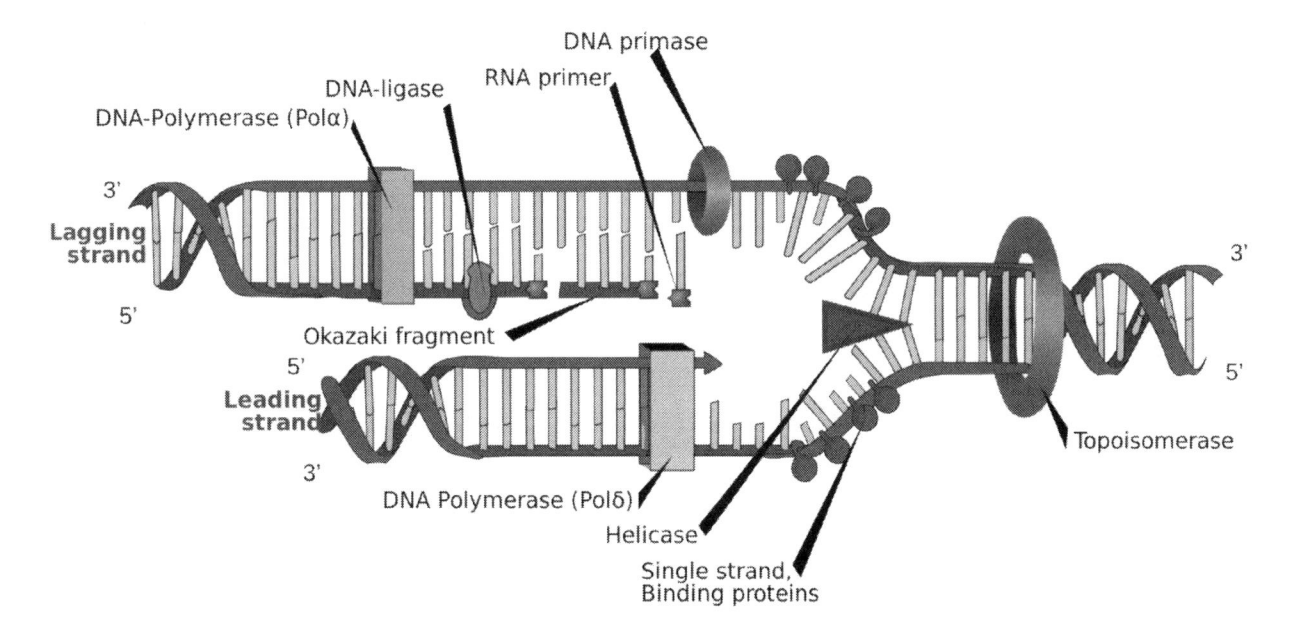

- DNA helicase unwinds the double strand to begin the process of DNA replication
    - This opens up a replication fork
    - Original strand serves as the template for synthesis of the new strand
- DNA is copied by DNA polymerase
    - Initiated by an RNA primer synthesized by RNA polymerase
    - The new DNA strand is synthesized in the 5' → 3' direction
    - Leading strand synthesized continuously
    - Lagging strand synthesized discontinuously
        - Results in Okazaki fragments
        - RNA primers of the lagging strand are removed by DNA polymerase
        - Okazaki fragments are joined together by DNA ligase

- DNA replication is semiconservative
    - Each new DNA molecule contains a single parental strand and one new strand
- DNA replication in some bacteria like E. coli is bidirectional around the chromosome
- DNA replication is very accurate; mistakes occur once in every 10^{10} nucleotides
    - DNA polymerase can excise any base that doesn't properly complement the parental DNA strand during synthesis and replace it with an appropriate base
    - This proofreading ability is the 3' to 5' exonuclease activity of DNA polymerase

RNA and Protein Synthesis

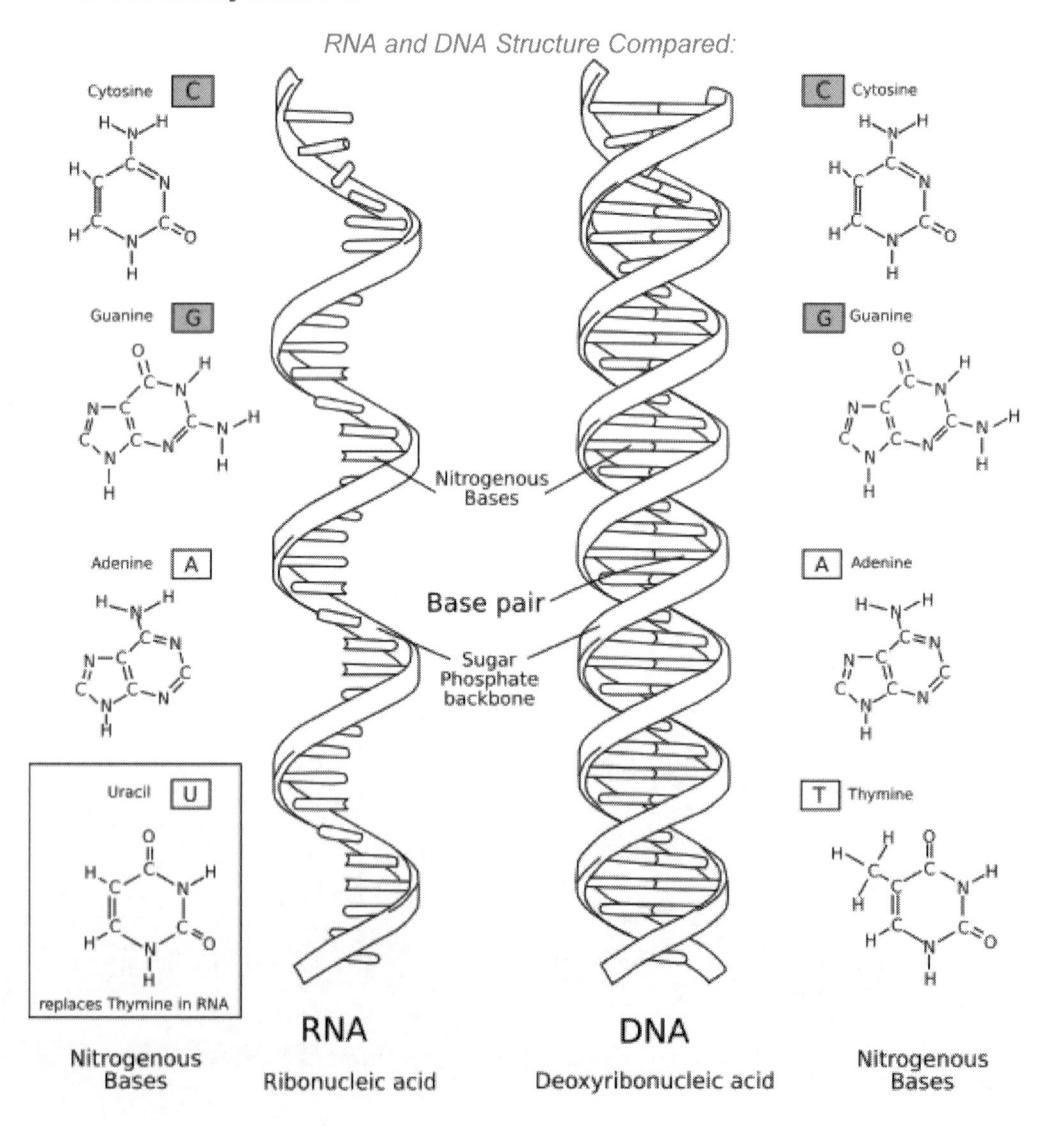

RNA and DNA Structure Compared:

Transcription

- DNA is transcribed to make different forms of RNA
    - Messenger RNA (mRNA)
    - Transfer RNA (tRNA)
    - Ribosomal RNA (rRNA)
- RNA is a single-stranded polymer of nucleotides
    - Adenine (A), uracil (U), cytosine (C), and guanine (G)
- RNA "backbone" is a repeated pattern of ribose, phosphate, ribose, phosphate...
- A strand of mRNA is synthesized using a single strand of DNA as a template
    - Usually a single gene on the DNA strand is used as a template for the synthesis of an mRNA molecule
    - Process is similar to DNA replication, except for the fact that there are no thymine (T) nucleotides in RNA
        - Instead, the nucleotide uracil (U) is present in place of thymine
- Transcription begins when RNA polymerase binds to the promoter sequence on the DNA template strand
- Synthesis of the RNA transcript proceeds in the 5' → 3' direction
- RNA polymerase assembles the mRNA strand from free nucleotides through complimentary base pairing using the DNA template strand as a guide
- Transcription stops when the RNA polymerase reaches the terminator sequence and the mRNA and the RNA polymerase are released from the DNA
- Process of transcription allows the cell to produce short-term copies of genes that can be used as the direct source of information for protein synthesis
    - Messenger RNA acts as an intermediate between the permanent storage form, DNA, and the process that uses the genetic information, translation

Translation

- mRNA is translated in blocks called codons, each of which consist of 3 nucleotides
    - The sequence of codons specifies the order in which amino acids will be assembled into proteins

- o Often, multiple codons specify the same amino acid
  - Referred to as the degeneracy of the code
  - Only 20 amino acids, but there are 64 possible codons

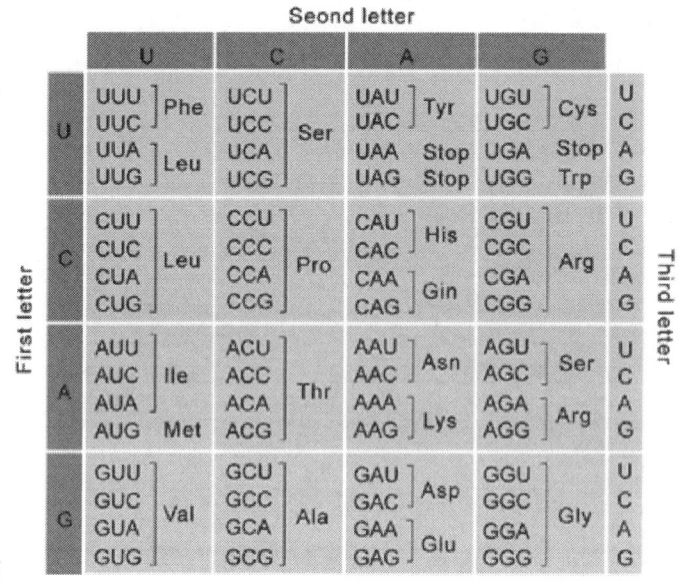

The Genetic Code:

- Translation of mRNA begins at the start codon AUG
- Translation ends at a STOP (nonsense) codon: UAA, UAG, UGA
- Translation is facilitated by the ribosome

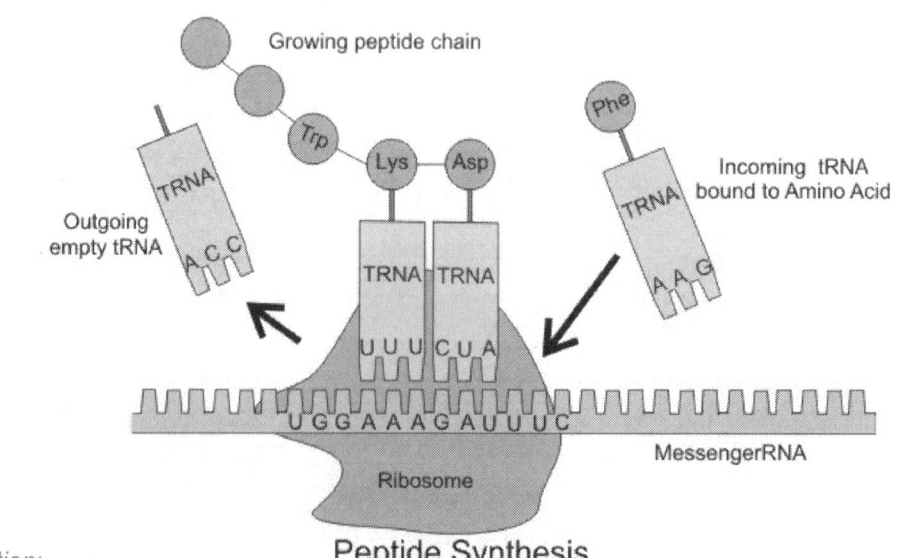

Translation: Peptide Synthesis

- tRNA molecules recognize the codons, and each carries a specific individual amino acid
    - Each tRNA has a region called an anticodon that is complimentary to a specific codon on the mRNA
    - On the opposite end of the tRNA a single specific amino acid is attached
    - Ribosome directs the orderly binding of tRNAs to the codons and assembles the polypeptide chain
    - Ribosome moves along the mRNA transcript in the 5' → 3' direction
    - Multiple ribosomes can be translating the same mRNA molecule at the same time
        - Allows a cell to produce large quantities of a protein
    - In prokaryotes, translation can begin even before transcription is complete

RNA processing in Eukaryotes

- Transcription occurs in the nucleus
- Genes of eukaryotic DNA contain introns and exons
    - Only exons code for proteins
- Small nuclear ribonucleoproteins (snRNPs) and ribozymes remove the introns from the mRNA transcript and splice together the exons
    - Ribozymes are like enzymes, but they are made of RNA and act on RNA
- RNA processing produces a functional mRNA molecule
- mRNA leaves the nucleus for translation in the cytoplasm

Regulation of Bacterial Gene Expression

Cells conserve energy by synthesizing only the proteins they need at the time.

- Constitutive genes (60-80% of all genes) are expressed continuously
    - Facultative genes are expressed only when needed

- Repression and induction regulate mRNA transcription
    - Repression inhibits transcription of DNA
        - Repressor proteins block RNA polymerase from initiating transcription of particular genes
    - Induction
        - Inducers turn on the transcription process of a gene

The *lac* Operon

The lac operon is required for the transport and metabolism of lactose in E.coli. An operon is a functioning unit of DNA that contains multiple genes transcribed from one promoter (operator region). All genes are transcribed together.

- Operon encodes for polycistronic mRNA
    - Contains the coding sequence for two or more genes
- Operon consists of a promoter, terminator, genes, operator
- DNA elements of the *lac* operon
    - Promoter region
        - Binds RNA polymerase
    - Operator region
        - Binds *lac* repressor protein
    - CAP site
        - Binds CAP (catabolite activator protein)
- Genes encoded by the *lac* operon
    - *lacZ* gene encodes β-galactosidase
        - Hydrolyzes lactose to galactose + glucose
        - Converts lactose into allolactose
    - *lacY* gene encodes lactose permease
        - Transporter for lactose and its analogues
    - *lacA* gene encodes thiogalactoside transacetylase
        - Function is unclear

- *lacI* gene encodes the *lac* repressor
    - Not part of *lac* operon
    - Blocks transcription
    - Repressor binds to operator
        - Blocks σ factor from binding promoter
    - Always present
        - Default expression is OFF for the *lac* operon
- *lac* operon can be regulated by:
    - Repressor protein
        - Inducible, negative control mechanism
        - Allolactose is the inducer
            - Binds to the repressor and inactivates it
    - Activator protein
        - Inducible, positive control mechanism
        - Small molecule involved is cAMP (cyclic AMP)
        - cAMP binds to CAP forming the cAMP-CAP complex
            - Complex binds to the CAP site and increases transcription
        - When glucose levels are high in the cell, cAMP levels are low
            - Means cAMP is not available to bind CAP
                - Transcription rate is decreased
- Mutants
    - I^- mutant – has a repressor that can't bind to the operator
        - *lac* operon is constitutively expressed
    - I^s mutant – has a repressor that can't be inactivated
        - No expression
    - O^C mutant – has an operator that the repressor can't bind to
        - *lac* operon is constitutively expressed

- *lacP*- mutant – RNA polymerase can't bind to the DNA
    - No expression

The *lac* Operon and its Control Elements

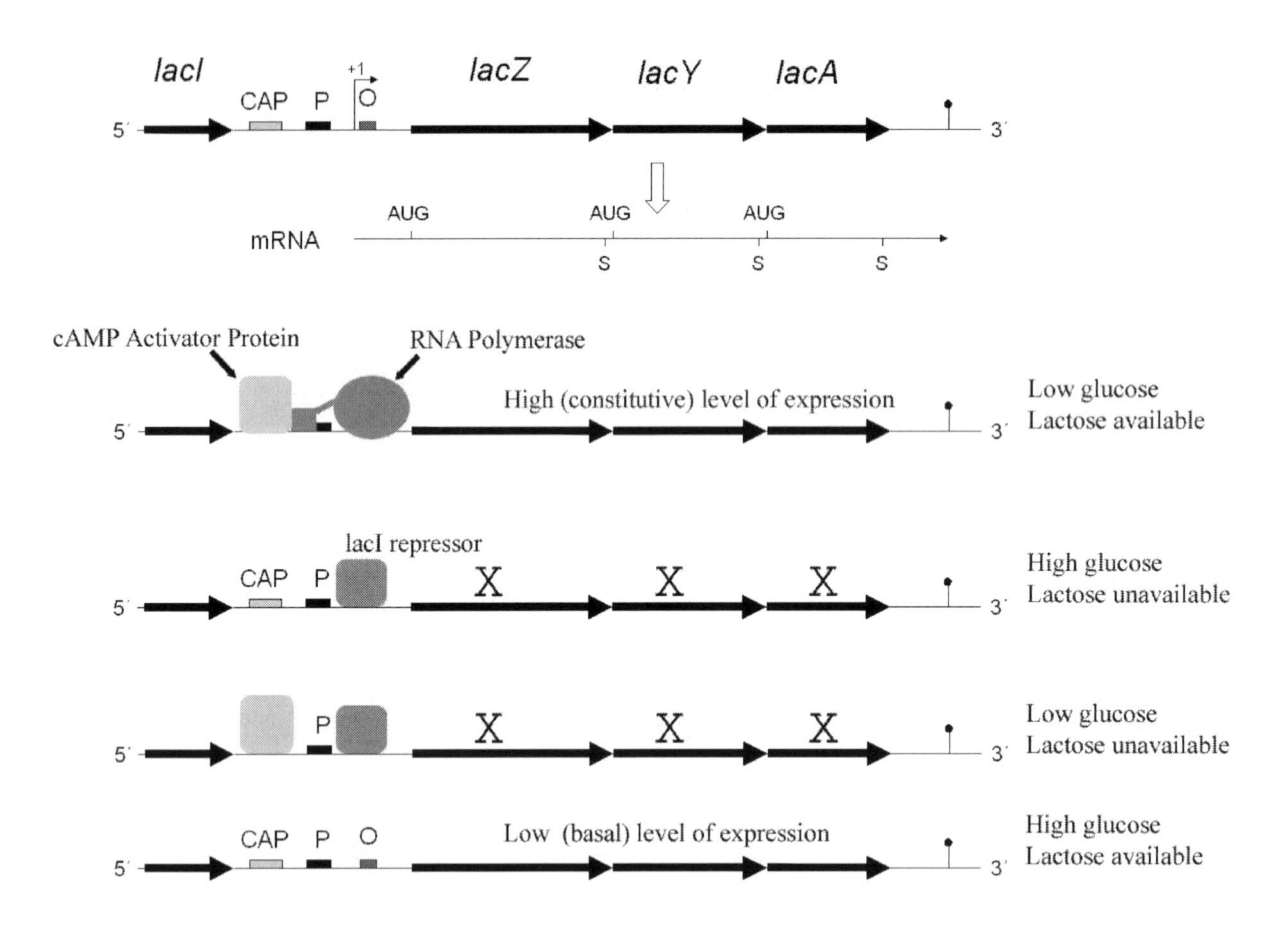

Mutation

The changing of the structure of a gene, resulting in a variant form that may be transmitted to subsequent generations, caused by the alteration of single base units in DNA, or the deletion, insertion, or rearrangement of larger sections of genes or chromosomes.

- Mutations that change the base sequence of DNA may be neutral (silent), beneficial (adaptive), or harmful (deleterious)
    - Neutral mutations neither benefit nor damage the organism
    - Adaptive mutations could lead to antibiotic resistance or altered pathogenicity
    - Deleterious mutations lead to loss of function, the formation of a cancer, or death

Types of Mutations

- Point mutation (substitution) – single nucleotide substitution
    - Two types
        - Transition - purine changed to another purine or pyrimidine changed to another pyrimidine
        - Transversion - purine changed to pyrimidine or vice versa
- Deletion – deletion of at least 1 nucleotide
- Insertion – insertion of at least 1 nucleotide
- Mistakes by DNA polymerase
    - Mis-matched bases
- Damage by ROS (Reactive Oxygen Species) like •O_2- or H_2O_2
    - By-products of oxidative metabolism
    - Example: G oxidized to oxoG
        - Can base pair with either C or A
- Spontaneous depurination
    - Glycosidic bond connecting base to sugar is broken
    - Results in abasic site
    - Occurs ~18,000X/day
- Spontaneous deamination
    - Removal of an amine group
    - Original G:C base pair can become T:A
    - Occurs ~500 times per day

Mutagens Cause Mutations

- Mutation Sources
    - Electromagnetic radiation
    - X-rays, gamma rays
    - UV light

The Frequency of Mutation

- Spontaneous mutation rate = 1 in 10^9 replicated bases
- Mutagens increase the mutation rate to one in every 10^3 to 10^5

Identifying Mutants

- Positive (direct) selection detects mutant cells because they grow or appear differently from the rest of the population, which lack the mutation
- Negative (indirect) selection detects mutant cells because they do not grow
  - Replica Plating
    - Mutant cells that have a nutritional requirement that is absent in the non-mutant parent cells are said to be auxotrophs

Genetic Transfer and Recombination

Genetic recombination

- Exchange of genes between two DNA molecules to form new combinations of genes on a chromosome
  - Crossing over in eukaryotes
- Increases a population's genetic diversity
- Recombination is more likely to produce a beneficial outcome than mutation
  - Recombination is less likely to destroy a gene's function
  - It may bring together a combination of genes that enable the organism to carry out a valuable new function

Vertical Gene Transfer

- When genes are passed from parent cell to daughter cell

Horizontal Gene Transfer

- When genes are passed from one adult cell to another
- Donor cell gives a portion of its DNA to a recipient cell
  - Part of the donor's DNA is incorporated into the recipient's DNA
    - Recipient cell is then called a recombinant cell
    - This is a very rare occurrence and may happen in less than 1% of a population

Transformation in Bacteria

- When a cell lyses, its DNA is released
- Closely related living cells can take up fragments of that DNA and incorporate it into their own DNA
    - This forms a hybrid recombinant cell
- All descendants of the recombinant will be identical to it
- Transformation naturally occurs only among a few genera of bacteria

Conjugation in Bacteria

Bacterial conjugation refers to the transfer of genetic material between bacterial cells by direct cell-cell contact or a bridge-like connection between the two cells.

- Only certain strains of bacteria can donate the genes they carry
    - Contain a plasmid carrying the F factor
    - Plasmid itself is called the F plasmid
        - Plasmid - small, circular piece of DNA that replicates independently from the cell's chromosome
            - Plasmids transferred during conjugation are called conjugative plasmids
        - Strains that have the F factor are referred to as being F^+
            - Cells lacking it are F^-
- In F+ cells, plasmids carry genes that code for the synthesis of a sex pili
    - Sex pili help bring two cells together for the transfer
- Bacterial conjugation steps
    - Conjugation is initiated between a F+ and F- cell
    - One strand of the F factor is cut by an endonuclease
        - Strand moves across the conjugation tube
    - DNA complement is synthesized on both single strands
    - Ligase seals both strands reforming the plasmid in both the cells

- Conjugates separate
  - Both cells are F⁺ at the end since they both contain the F plasmid

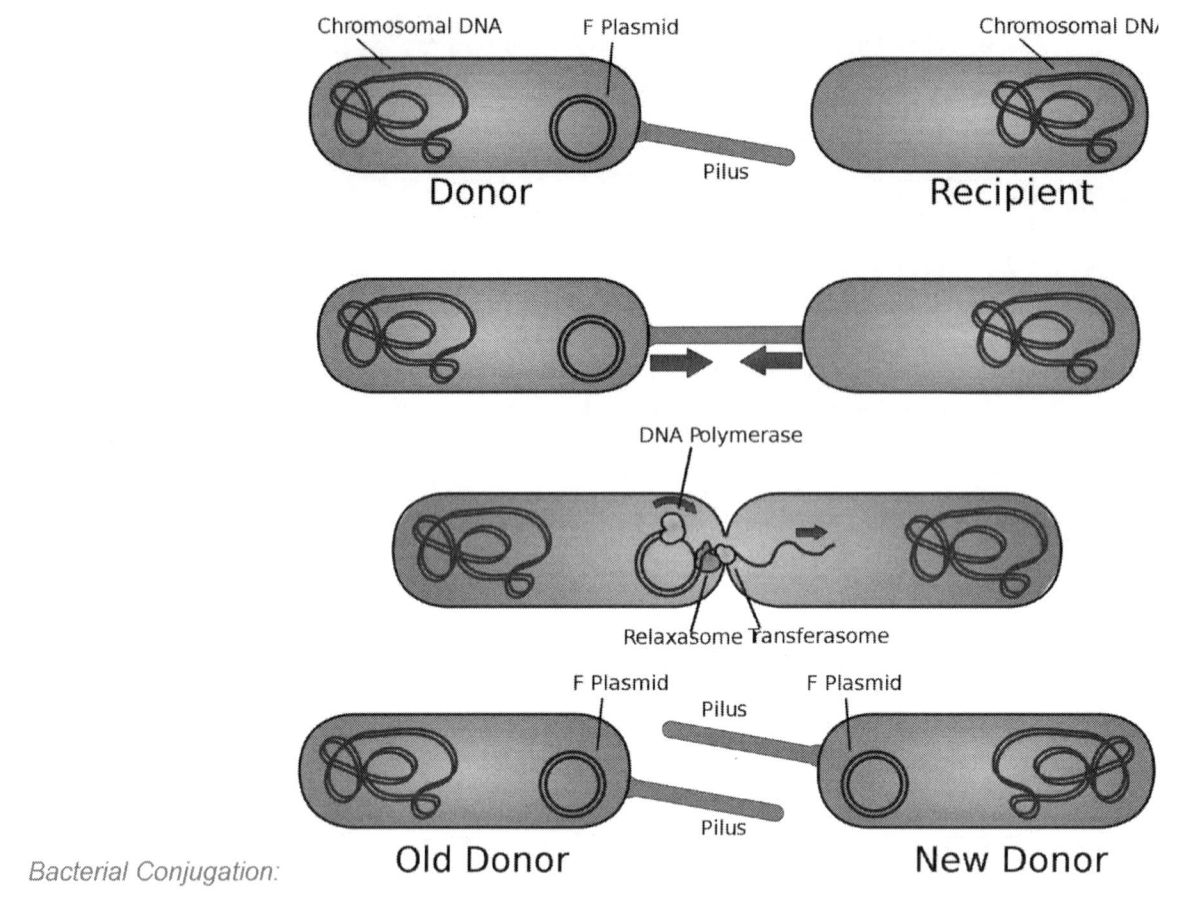

Bacterial Conjugation:

Transduction in Bacteria

Transfer of genetic material using a genetic vector. The vector is a bacteriophage or simply a phage, which are viruses that infect bacteria.

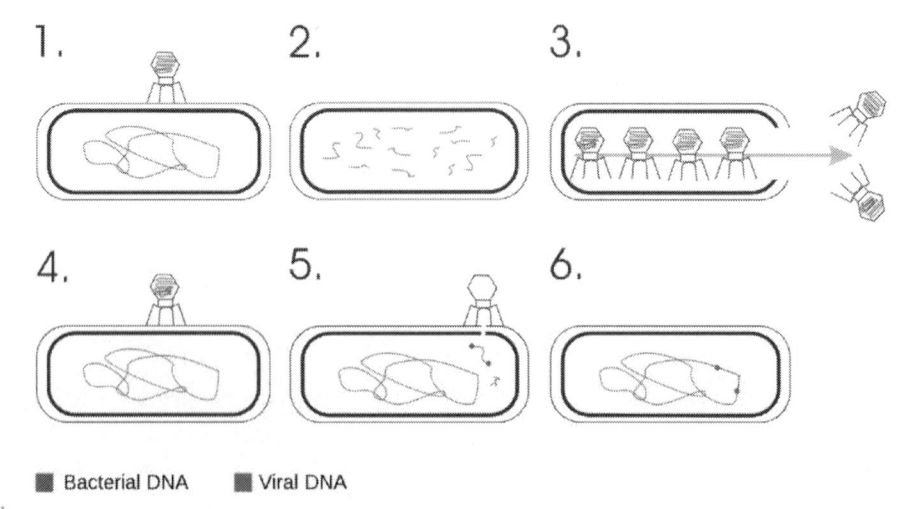

Transduction:

Transduction Steps

- Step 1: Phage latches on to a bacterium
- Step 2: Phage injects its genetic material into the cell
    - Phage DNA replication occurs and host genome is fragmented
- Step 3: Packaging of replicated phage DNA and synthesis of new phages
    - Some of the host cell DNA is packaged with the viral DNA
    - Cell lyses open and releases mature phage into its surroundings
- Step 4: Phage carrying donor host DNA latches onto another cell
- Step 5: Introduction of donor host DNA into recipient cell
- Step 6: DNA from the first bacterial host is recombined into the chromosome of the new host

Plasmids and Transposons

Plasmids and transposons are genetic elements that provide additional mechanisms for genetic change that occur in both prokaryotes and some eukaryotes.

Plasmids

- Conjugative plasmid - carries genes for sex pili and transfer of the plasmid
- Dissimilation plasmids - encode enzymes for catabolism of unusual compounds
- Resistance (R) factors - encode resistance to antibiotics, heavy metals, or cellular toxins
    - Treatment with antibiotics has selected for the survival of bacteria that have R factors
    - Transfer of resistance contributes to the problem of antibiotic resistant bacteria

Transposons

Transposons are segments of DNA that can move from one region of a DNA molecule to another, from one chromosome to another, or from a chromosome to a plasmid.

- Contain insertion sequences for cutting and resealing DNA (transposase)
    - If insertion occurs within a gene, that gene will be rendered nonfunctional
    - Frequency of transposition is about the same as the spontaneous mutation rate of DNA in bacteria, it a very rare event
- Complex transposons carry other genes in addition to the insertion sequences
- Can be carried between cells on plasmids or in viruses
    - Can therefore be spread between members of the same species, and between members of different species

Genes and Evolution

Genetic variation resulting from mutation, vertical gene transfer (crossing over), horizontal gene transfer (transformation, conjugation, transduction, and transposition) drive evolution via natural selection.

CHAPTER 9: BIOTECHNOLOGY AND RECOMBINANT DNA

Introduction to Biotechnology

Biotechnology is the use and manipulation of microorganisms, cells, or cell components to make desired products (antibiotics, vitamins, etc.).

- Genetic engineering - process of using recombinant DNA technology to create new cells that produce chemicals that an organism doesn't naturally make

- Recombinant DNA technology - insertion or modification of genes in an organism to produce desired proteins/enzymes

    o Genes from one organism can placed inside another organism's DNA

        ▪ Including organisms of different species

    o Examples:

        ▪ Human gene for insulin production has been inserted into bacteria and they are able to produce human insulin for commercial use

        ▪ A gene that codes for a viral coat protein from the hepatitis B virus has been inserted into yeast for the commercial production of a vaccine against this disease

Overview of Recombinant DNA Procedures

- Gene of interest is inserted into a self-replicating vector (plasmid, transposon, viral DNA)

    o Accomplished through the use of restriction enzymes which can cleave DNA molecules at restriction sites

- Recombinant vector is then taken up by a cell

- The cell is then allowed to divide producing a culture
    - All cells in the culture will be identical
        - Each carries the vector with the gene of interest
    - The gene of interest can be isolated in large quantities for further experimentation
    - Expression of the gene of interest within the cells can produce a large volume of product (protein, enzyme, hormone, etc.) that can then be harvested

Tools of Biotechnology

- Selection - culturing of a naturally-occurring microbe that produces a desired product
- Site-directed mutagenesis - changing a specific gene to change a protein

Restriction Enzymes

- Cut specific sequences of DNA
- Naturally occur in some species of bacteria
    - Destroy bacteriophage (viral) DNA that gets into these species' cells
    - Cannot digest the bacterial (host) DNA because of the presence of methylated cytosines
- Purified forms of bacterial restriction enzymes are used in genetic engineering
- A specific restriction enzyme always recognizes and cuts DNA at a very specific nucleotide sequence of the DNA molecule
- Cuts made by some restriction enzymes are staggered, producing "sticky ends"
    - Cuts that are made, are the same on both strands of the DNA
    - However, they run in opposite directions

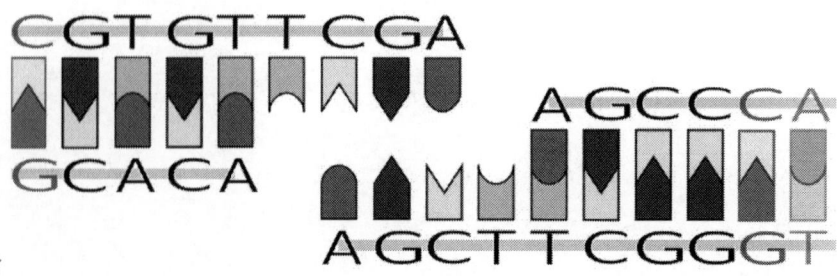

Example of "Sticky Ends":

- o Sticky ends can hydrogen bond with a complimentary base sequence
- o If two fragments of DNA from different sources have been cut with the same restriction enzyme, the two fragments will have complimentary sticky ends and can recombine
- DNA ligase can then be used to covalently link the sugar-phosphate backbones
  - o Produces a recombinant DNA molecule

Vectors

- Transport DNA into the desired cell
- Plasmids, transposons, and viruses are potential vectors
- Characteristics of good vectors
  - o Must be self-replicating
  - o Should resist destruction by the recipient cell
    - Circular vectors are highly resilient
    - Viral DNA that inserts itself quickly into the host chromosome is more likely to remain intact
  - o Should carry a marker gene
    - Makes it easy to retrieve clones containing the vector
    - For example, a marker gene may code for the production of a specific enzyme, or antibiotic resistance
- Shuttle vectors are plasmids capable of existing in several different species
  - o Used to move genes from one species to another
- Vectors can be used to insert functional genes into human cells that have defective genes
  - o This is gene therapy

Polymerase Chain Reaction (PCR)
- Used to make multiple copies of a piece of DNA
- Also used to:
    - Amplify DNA for recombination experiments
    - Sequence DNA
    - Diagnose genetic diseases
    - Detect pathogens

PCR Steps
- DNA of interest, RNA primers, DNA nucleotides, and DNA polymerase are placed in a thermocycler
- Contents are incubated at 94°C for a minute
    - Strands of the original piece of DNA separates
- Temperature is then lowered to 60°C for a minute
    - RNA primers anneal to the DNA
- Temperature is then increased to 72°C for a minute
    - DNA polymerase synthesizes a complimentary strand of DNA
- Cycle is repeated several times, resulting in an exponential increase in the number of DNA pieces
    - All of which are identical to the original piece of DNA
- DNA polymerase from the thermophilic bacterium *Thermus aquaticus* is used because of the high temperatures involved
    - This polymerase is able to function at high temperatures which would denature normal polymerases
- PCR can only be used to amplify small specific sequences of DNA as determined by the primer used.
    - Cannot be used to amplify an entire genome

Techniques of Genetic Engineering

Recombinant DNA in the lab is made outside a living cell. Once the recombinant DNA has been made, it must be put back into a cell. DNA can be inserted into a cell by:

Inserting DNA into a Cell

- Transformation
    - Some species spontaneously take up a recombinant plasmid and integrate it into their chromosomes by recombination
    - Many species are unable to spontaneously transform
        - They first must be treated with chemicals to make them "competent" (i.e. able to take up external DNA)
- Electroporation
    - Application of an electrical current to a cell forms small pores in the plasma membrane that the DNA can pass through
        - Cells with walls must first be converted to protoplasts or spheroplasts (Gram+ and Gram- cells, respectively, without their cell walls) for this technique to work
- Protoplast fusion
    - Protoplasts in solution will fuse to form hybrid cells at a slow rate
        - Adding polyethylene glycol increases the rate of fusion
    - Once fusion occurs, the two chromosomes may undergo natural recombination
- Gene gun
    - Microscopic particles of gold or tungsten are coated with DNA and shot out of a gene gun with a burst of helium
- Microinjection
    - A glass micropipette is used to inject DNA into a cell

Obtaining DNA

- Two main sources of genes:
    - Gene libraries are made up of pieces of an entire genome stored in bacterial plasmids or in phages
    - Synthetic DNA is made by a DNA synthesis machine

Cloning the Genes of Eukaryotes

- Genes containing introns are not useful for genetic engineering
    - Makes them too large to work with
    - Introns will not be removed from the mRNA transcript in bacteria and the product will be non-functional
- Complementary DNA (cDNA) is made from an mRNA template by the enzyme reverse transcriptase
    - Result is a DNA molecule without introns
    - Commonly used to obtain eukaryotic genes that can then be inserted into a prokaryote

Selecting for Cells

Out of millions of cells, only a few may have successfully taken up the gene of interest.

Blue-White Screening

- Use a plasmid vector that contains a gene that codes for a chosen antibiotic resistance with the gene of interest
    - Host bacterium cannot survive on a medium that contains the antibiotic unless it has taken up the plasmid
- Plasmid vector has a second gene that codes for the enzyme Beta-galactosidase
    - Enzyme cleaves lactose into glucose and galactose
    - Beta-galactosidase gene has several sites that can be cut by restriction enzymes
        - Plasmid vector and foreign DNA containing the gene of interest are digested with the same restriction enzyme
    - Fragments of the foreign DNA may insert into the Beta-galactosidase gene
        - Renders it non-functional
- Recombinant plasmid is inserted into the antibiotic-sensitive bacteria by transformation

- Recombinant bacteria are grown on a medium called X-gal, which contains the chosen antibiotic, and a substrate for Beta-galactosidase
    - Only cells that contain the plasmid vector will grow on this medium
        - Since they have the gene for resistance to the chosen antibiotic
    - If foreign DNA was not successfully inserted into the Beta-galactosidase gene, the gene will code for functional enzyme that will hydrolyze a component of the X-gal medium to produce a blue colored compound
        - Produces blue colonies when a bacterium is allowed to divide
        - Blue color signals that these cells are non-recombinant
    - If foreign DNA was successfully inserted into the Beta-galactosidase gene, it will be non-functional and colonies of these cells will appear white on the X-gal medium
- Even with the successful production of recombinant cells, it is still not known if the desired gene of interest was inserted into the Beta-galactosidase gene, or whether some other DNA fragment was inserted (further testing is therefore required)
    - If the gene of interest codes for the production of an identifiable product, the bacterial isolate only needs to be grown in culture and tested
    - Otherwise, the gene itself must be identified in the bacterium via a procedure known as colony hybridization

Colony Hybridization

- DNA probes are short sequences of single-stranded DNA that are complimentary to the desired gene
    - DNA probes are synthesized in the lab
        - Some contain fluorescent dye
        - Some contain radioactive phosphorous
- DNA probes bind with the gene of interest and serve as markers for colonies that contain the gene of interest

Applications of Genetic Engineering

Therapeutic Applications

- Subunit vaccines
    - Nonpathogenic viruses carrying genes for a pathogen's antigens
- Gene therapy is used to replace defective or missing genes

Human Genome Project

- About 3 billion nucleotides make up the DNA of a typical adult human cell
- These nucleotides have been sequenced and all genes have been mapped
- May provide diagnostics and treatments by determining all possible proteins that can be produced

Sequencing Genomes

- Random Shotgun Sequencing
    - Small fragments of an organism's genome are sequenced
    - Then a computer is used to assemble the sequences in the proper order
- Computer software exists that can find the protein encoding regions of the sequenced DNA
    - Which can then be isolated and used in additional recombinant experiments
- Computer assisted DNA sequencing has led to a new field of study, bioinformatics (science of studying how genes function)
    - Proteomics is the science of determining all of the proteins a cell can express

Genetic Screening

- Recombinant DNA technology is also used in genetic testing for the presence of genetic diseases
- Southern blotting is a technique that can be used in the genetic screening process
    - DNA containing the gene of interest is removed from the cell and cut into pieces with restriction enzymes
    - Fragments are separated using gel electrophoresis

- Fragments are then transferred to a filter by blotting
    - NaOH is used to separate the DNA fragments into single stranded molecules
    - Fragments on the filter are then exposed to a radioactive probe made from the defective form of the cloned gene of interest
    - Probe will hybridize with the defective gene if it is present in the sample of DNA taken from the cell
- Radioactive probes that form hybrids with the defective gene are detected by exposing the filter to X-ray film
- This technique can be used to test any person's DNA for the presence of a known defective gene
    - There are several hundred known defective genes that cause genetic diseases

Agricultural Applications

- Increase crop yields
- Make the plants resistant to herbicides, pests, drought, frost, viral infection, etc.
- Make the plants produce insecticides that kill insect predators
- Increase the shelf-life of fruits and vegetables after harvest
- Fix atmospheric N_2, to reduce the amount of fertilizer that is applied to crops

CHAPTER 10: CLASSIFICATION OF MICROBES

The Study of Phylogenetic Relationships

- Taxonomy - science of classifying organisms, based on common physical characteristics
    - Provides universal names for organisms
    - Provides a reference for identifying organisms
- Systematics or phylogeny
    - Study of the evolutionary history of organisms
- Phylogenetic Hierarchy
    - Descent from a common ancestor
        - Based on common properties in genetics, fossil structure, etc.

Classification of Organisms

- Taxonomic Hierarchy
    - Domain, Kingdom, Phylum (Division), Class, Order, Family, Genus, Species
- Classification of Prokaryotes
    - Based on rRNA sequence similarities
    - Clone - population of cells derived from a single cell
    - Strain - genetically different cells within a clone
- Classification of Eukaryotes
    - Domain Eukarya
        - Kingdom Animalia: multicellular; no cell walls; chemoheterotrophic
        - Kingdom Plantae: multicellular; cellulose cell walls; usually photoautotrophic

- Kingdom Fungi: chemoheterotrophic; unicellular or multicellular; cell walls of chitin; develop from spores or hyphal fragments
- Kingdom Protista: eukaryotic unicellular and multicellular organisms that do not fit in other kingdoms

- Classification of Viruses
    - Viral species - population of viruses with similar characteristics that occupies a particular ecological niche

CHAPTER 11: VIRUSES

General Characteristics of Viruses

- Inert when not within a host cell, require a living host cell to reproduce
    - Considered to be obligate intracellular parasites
- Viruses contain DNA or RNA, but not both
- No ATP generating mechanism
- Viruses have a protein coat surrounding the nucleic acid
    - Some viruses are also enclosed by an outer lipid envelope
- Little or no metabolic activity
- Viruses multiply inside living cells by using the host cell's own enzymes, nucleic acids, amino acids, ATP, tRNAs, ribosomes, etc.
    - Most drugs that would interfere with viral multiplication would have the same effect on the host cell

Host Range

- Host range - range of host cells a virus can infect
- Determined by specific host attachment sites and cellular factors necessary for viral multiplication
    - Attachment sites include things such as cell walls, fimbriae, flagella, and plasma membrane proteins
- Most viruses infect only specific types of cells in one host species
    - Bacterial viruses are called bacteriophages or phages

Viral Size

- Range from 20 to 1000 nm in length

Viral Structure

A virion is a single mature, complete, and infectious viral particle.

Nucleic Acid

- DNA or RNA is the primary genetic material in a virus
- Can be either a single stranded nucleic acid or double stranded nucleic acid
    - Single strand DNA (ssDNA)
    - Double strand DNA (dsDNA)
    - Single strand RNA (ssRNA)
    - Double strand RNA (dsRNA)
- Nucleic acid can be linear or circular
    - In some viruses, the nucleic acid is in several separate segments

Capsid and Envelope

- Capsid - protein coat surrounding the nucleic acid of a virus
    - Composed of subunits called capsomeres
        - Proteins composing the capsomeres may all be the same, or may be composed of several different proteins
- Envelope - lipid layer external to the capsid
    - Also contains proteins and carbohydrates
        - Some animal viruses become coated with the host cell's plasma membrane when released
    - May or may not be covered with glycoproteins (antigens) that project from the surface of the envelope
        - In some viruses, these glycoproteins serve as attachment points for binding to host cells
    - Viruses that have an envelope are simply called "enveloped viruses"
    - Nonenveloped viruses do not have an envelope

- Infection with a virus stimulates the host to produce antibodies that will recognize and bind to viral surface proteins (antigens)
    - Some viral proteins frequently mutate
        - Allows viral strains to infect a host more than once

General Morphology

Viruses can be classified on the basis of their structure.

- Helical
- Polyhedral (many-sided)
- Enveloped
- Complex

Taxonomy of Viruses

- Viruses are grouped into families based on:
    - Nucleic acid type
    - Replication strategy
    - Morphology
- Family names end in –viridae
- Genus names end in –virus
- Viral subspecies are designated by a number

Example

- Family: Herpesviridae
- Genus: Simplexvirus
- Common name: Human herpes virus (HHV)
- Subspecies: HHV 1, HHV 2, HHV 3

Biosynthesis and Multiplication of Viruses

Ribosomes, tRNA, amino acids, ATP, and most enzymes needed for protein and nucleic acid synthesis are supplied by the host cell and are used to synthesize viral proteins, enzymes, and nucleic acids.

Bacteriophage Multiplication

Bacteriophages are DNA viruses that multiply by two alternative mechanisms: the lytic cycle and the lysogenic cycle.

- Lytic Cycle
    - Phage attaches by tail fibers to the host cell
        - Viral attachment site binds to a complimentary receptor site (protein/antigen) on the bacterium
    - Phage lysozyme opens the cell wall and its tail sheath contracts to insert the tail core and viral DNA into the cell
    - Phage DNA circularizes and directs production of phage DNA and proteins
        - Host DNA is fragmented
    - Maturation - spontaneous assembly of phage components into virions
    - Phage lysozyme breaks cell wall and lyses open the host cell to release new virions
- Lysogenic Cycle
    - Lysogenic phages may proceed through a lytic cycle, but they can also insert their DNA into the host cell's DNA and begin a lysogenic cycle
    - In lysogeny, the phage remains inactive
        - Host cell remains alive during lysogeny and is called a lysogenic cell
    - Phage DNA recombines with host DNA to form a prophage
        - Most prophage genes are repressed by repressor proteins
        - Lysogenic bacterium reproduces normally
        - Occasionally, the prophage may excise itself from the bacterial chromosome to initiate the lytic cycle

Multiplication of Animal Viruses

- Virus attaches to proteins/glycoproteins of the plasma membrane
- Endocytosis encloses the virion within a vesicle
    - If the virus is enveloped, the envelope is destroyed when lysosomes fuse with the vesicle
    - Capsid may also be digested or it may be released into the cytoplasm of the host
        - Lysosomes do not contain enzymes to break apart DNA or RNA
- Fusion is an alternate method by which enveloped viruses can enter the host cell
    - Viral envelope fuses with the plasma membrane of the host, releasing the capsid into the cell's cytoplasm
        - Uncoating - separation of the viral nucleic acid from the capsid via viral or host enzymes
    - Biosynthesis - production of viral nucleic acid and proteins
    - Maturation - nucleic acid and capsid proteins spontaneously assemble
- Budding occurs with the release of enveloped viruses
    - Envelope is partially derived from the host plasma membrane or nuclear membrane
    - Does not necessarily kill the cell
- Nonenveloped viruses are released when the host's plasma membrane ruptures (host dies)

Biosynthesis of DNA Animal Viruses

- Attachment, penetration, and uncoating occur
- Viral DNA enters the host nucleus, and one of two things can happen:
    - Viral DNA can recombine with the host DNA to form a provirus
        - Provirus remains permanently within the host DNA
            - Different from a prophage, which can excise itself from the host chromosome

- - - Provirus may stay inactive within the host, but it will be replicated whenever the host DNA replicates
      - Mutagens can induce expression of a provirus that was previously latent
  - Alternatively, the viral DNA/provirus can be expressed, producing new virions
    - "Early genes" of the virus are transcribed and translated to produce proteins needed for viral DNA replication in the nucleus
    - Viral DNA replication then occurs using the host's DNA polymerase within the nucleus
    - "Late genes" of the virus are transcribed and translated to produce capsid proteins and other structural proteins
    - Capsid proteins migrate into the nucleus where maturation occurs
    - Virions are then transported along the endoplasmic reticulum to the host cell's membrane for release via budding or rupturing
- New virions are released

Biosynthesis of RNA Animal Viruses

Nearly the same processes as seen in the DNA viruses, except for the ways in which viral mRNA and viral RNA are produced.

- Biosynthesis of Positive-Strand RNA Animal Viruses
  - Attachment, penetration, and uncoating occur
  - Viral RNA is translated into two principal proteins
    - One inhibits host cell synthesis of RNA and protein
    - Other is RNA-dependent RNA polymerase, which catalyzes the synthesis of another strand of viral RNA, complementary to the original positive viral RNA strand
      - New strand is an antisense (negative) strand, which serves as a template to produce additional positive strands
  - The positive RNA strands serves as mRNA for the translation of viral proteins

- Once viral RNAs and viral proteins are synthesized, maturation and release occur
    - Negative strand RNA molecules are broken down
- Biosynthesis of Negative-Strand RNA Animal Viruses
    - Exactly the same as biosynthesis of positive-strand RNA animal viruses, except, instead of negative strand RNA molecules being broken down, the positive strand RNA molecules are broken down in this version

Multiplication of a Retrovirus in Animals

- Attachment, penetration, and uncoating occur
- Members of this viral family contain two identical strands of positive RNA and the enzyme reverse transcriptase
    - Reverse transcriptase uses RNA as a template to make a new complimentary DNA (cDNA) strand
- Newly synthesized cDNA strand is then used as a template for the synthesis of another complimentary DNA strand, antiparallel to the cDNA, yielding a double-stranded DNA molecule
    - Reverse transcriptase also destroys the original viral RNA
- Viral DNA recombines with the host DNA to form a provirus
- In order to produce new virions; the viral DNA must be transcribed back into mRNA that will direct viral protein synthesis and be incorporated into new virions
    - Viral mRNA is transcribed in the nucleus and transported into the cytoplasm
    - Translation occurs in the cytoplasm to produce capsid proteins and other structural proteins
    - Maturation occurs in the cytoplasm
    - New virions are then released by budding (for enveloped viruses) or rupturing (for non-enveloped viruses)

Viruses and Cancer

Cancer is the uncontrolled growth of cells. Most cancers are thought to involve multiple mutations that accumulate sequentially.

- Mutation in a cell can disrupt its normal cell cycle causing it to reproduce faster than normal cells
- Among the mutant and its progeny, one cell may undergo another mutation in another gene, causing even more disruption in the cell cycle
    - Disruption of the cell cycle leads to faster mutant cell reproduction which leads to increasing rounds of DNA replication (additional chances for more mutations to occur)
- Formation of a provirus represents a permanent genetic change in a cell
    - Viruses that can form proviruses can cause cancer
- Oncogene - a gene that causes cancer (increases the rate of cellular reproduction) when it experiences mutation
    - Activated oncogenes transform normal cells into cancerous cells
- Tumor cells undergo transformation to acquire new properties
    - Become dedifferentiated (less specialized than normal cells)
    - Have an increased growth rate
    - Lose contact inhibition
    - Become transplantable (can detach from one tissue and re-attach to another tissue = metastasis)
    - Become invasive (squeeze into tight spaces that normal body cells can't)
- Genetic material of oncogenic viruses becomes integrated into the host cell's DNA
    - Two types of oncogenic viruses
        - DNA oncogenic viruses
        - RNA oncogenic viruses

Latent vs. Persistent Viral Infections

Latent

- Virus remains in an asymptomatic host cell for a long period of time
    - Then suddenly becomes active to cause disease
- Trigger for disease manifestation is usually a stressor
    - Infection with another microbe, injury, psychological stress, etc.

Persistent

- Disease progress occurs slowly over a long period of time, generally fatal

Prions

- Infectious proteins that are inherited and transmissible by ingestion, surgical transplant, and contaminated surgical instruments
- PrP^C, normal cellular prion protein, found on the surface of neurons
- PrP^{Sc}, scrapie protein, accumulate in brain cells forming plaques

Plant Viruses and Viroids

Plant viruses must enter through a wound in the plant or gain entry with the help of a plant parasite (nematode, fungus, insect). Infected plants can spread the virus to other plants via pollen and seeds.

- Viroids are naked RNA molecules between 300-400 nucleotides long that fold back on themselves.
    - RNA does not code for proteins
    - Strictly plant pathogens
    - May have evolved from introns

CHAPTER 12: PRINCIPLES OF DISEASE AND EPIDEMIOLOGY

Terminology

- Pathology - scientific study of disease
- Etiology - study of the cause of a disease
- Pathogenesis - development of a disease
- Infection - colonization of the body by pathogens, or the presence of a microbe in a part of the body where it is not normally found
- Disease - an abnormal state in which the body is not functioning normally

Normal Microbiota

The human body contains more bacteria cells than human cells.

- Normal microbiota are only found in certain regions of the body
- Transient microbiota may be present for days, weeks, or months
    - Normal microbiota are permanent

Relationships between Normal Microbiota and Host

- Microbial antagonism is competition between microbes
    - Normal microbiota prevent harmful microorganisms from overgrowing in the following ways:
        - Occupying niches that pathogens might occupy
        - Producing bacteriocins (compounds that kill other bacteria)
- Probiotics are live microbes applied to or ingested into the body, intended to exert a beneficial effect

Terminology

- Symbiosis - relationship between a microbe and its host
- Commensalism - relationship in which one organism benefits and the other is unaffected

- Mutualism – relationship in which both organisms benefit
    - Many of the normal microbiota are mutualists
- Parasitism, one organism is benefited at the expense of the other
    - Many disease causing organisms are parasites

Opportunistic Microorganisms

- Some normal microbiota are opportunistic pathogens
    - Normally do not cause disease in healthy individuals
    - Can cause disease if they colonize an area where they are not normally found
    - Individuals with compromised immune systems often become infected with opportunistic pathogens

Causes of Infectious Diseases

Koch's Postulates

- Used to prove the cause of an infectious disease
    - Same pathogen must be present in every case of the disease
    - Pathogen must be isolated from the diseased host and grown in pure culture
    - Pathogen from the pure culture must cause the disease when it is inoculated into a healthy and susceptible laboratory animal
    - Pathogen must be isolated from the inoculated animal and must be shown to be the original organism

Exceptions to Koch's Postulates

- Some microorganisms cannot be cultured on artificial media
- Multiple organisms can cause the same disease signs and symptoms
- Some species can cause multiple different disease conditions

Classifying Infectious Diseases (Terminology)

- Symptom - subjective change in body function that is felt by a patient as a result of disease
- Sign - an objective change in the body that can be measured or observed as a result of disease

- Syndrome - specific group of signs and symptoms that always accompany a disease
- Communicable disease - disease that is directly or indirectly spread from one host to another
- Contagious disease - disease that is easily spread from one host to another
- Non-communicable disease - disease that is not transmitted from one host to another

Occurrence of Disease

- Incidence - fraction of a population that contracts a disease during a specific time period
- Prevalence - fraction of a population having a specific disease at a given time
- Sporadic disease - disease that occurs only occasionally in a population
- Endemic disease - disease that is constantly present in a population
- Epidemic disease – disease acquired by many hosts in a given area in a short time
- Pandemic disease - worldwide epidemic

Severity or Duration of a Disease

- Acute disease – disease with a rapid onset
- Chronic disease - develops slowly and persists for long periods
- Subacute disease - intermediate between acute and chronic
- Latent disease - disease that does not display symptoms for a period of time
    - Disease causing organism is inactive within the host during this time
- Herd immunity - when most of a population is immune to a pathogen, non-immune individuals are less likely to come into contact with an individual that is infected

Extent of Host Involvement

- Local infection - pathogens limited to a small area of the body
- Systemic (generalized) infection - microorganisms or their products are spread throughout the body via blood or lymph
- Focal infection - systemic infection that began as a local infection

- Bacteremia - bacteria in the blood
- Septicemia - growth of bacteria in the blood
- Toxemia - toxins in the blood
- Viremia - viruses in the blood
- Primary infection - acute infection that causes the initial illness
- Secondary infection - opportunistic infection after a primary (predisposing) infection
- Sub-clinical infection - no noticeable signs or symptoms (inapparent infection)

Patterns of Disease

There is normally a specific sequence of events associated with infection by a microbe and subsequent disease.

- Reservoir (where an infectious agent normally lives and multiplies) must exist
- Pathogen must be transmitted to a susceptible host from the reservoir
- Pathogen invades the host and multiplies
- Pathogen harms the host
- Illness ends when:
    - Host dies
    - Or when the host's immune system destroys the pathogen

Predisposing Factors

Predisposing factors make the body more susceptible to disease.

- Short urethra in females
- Inherited traits (e.g. sickle-cell gene)
- Climate and weather
- Fatigue

- Age
- Lifestyle/nutrition
- Chemotherapy
- Radiation treatments

Development of Disease

The specific sequence of events that occurs when a pathogen establishes itself in the host:

- Incubation Period
    - Time between initial infection and the appearance of signs or symptoms
- Prodromal period
    - Early mild symptoms of the disease appear
    - Not all diseases have a prodromal period
- Period of Illness
    - Disease is most acute at this time
    - Overt signs and symptoms appear
    - Host's immune response usually overcomes the pathogen to end the period of illness
        - If not, the host dies
- Period of Decline
    - Signs and symptoms disappear
- Period of Convalescence
    - Host returns to normal

The Spread of Infection

Reservoirs of Infection

- Reservoirs of infection are continual sources of infection
    - Can be living or nonliving
- Human Reservoirs
    - Carriers may not have apparent infections or latent infections

- Animal Reservoirs
  - Zoonotic diseases are diseases transmitted to humans
- Nonliving Reservoirs
  - Soil, water, etc.

Transmission of Disease

- Direct contact transmission - requires close association between an infected individual and a susceptible host
  - Also known as person-to-person transmission
  - Pathogens can also be passed from animals to humans through direct contact
- Indirect contact transmission - pathogen is spread from its reservoir to a susceptible host via an inanimate object (a fomite)
- Droplet transmission - transmission via airborne droplets that travel less than one meter from the reservoir to the new host
- Vehicle-borne transmission - transmission by an inanimate reservoir that aren't fomites
- Waterborne transmission - normally a result of water being contaminated with sewage
- Foodborne transmission - normally the result of foods being undercooked, poorly stored/refrigerated, or handled in unsanitary ways
- Airborne transmission - transmission by airborne droplets that travel more than one meter from the reservoir to the new host
- Mechanical transmission - arthropods carry pathogens on their feet or other body parts
- Biological transmission - pathogen reproduces inside the vector

Nosocomial (Hospital-Acquired) Infections

- 5-15% of all hospital patients acquire nosocomial infections
- Some normal microbiota are opportunistic pathogens

- Some nosocomial pathogens are antibiotic resistant
- Compromised host - a host whose resistance to infection is impaired by disease, therapy, or burns
    - Broken skin, broken mucous membranes, or a suppressed immune system are all conditions that can compromise the host

Epidemiology

Epidemiology is the study of where and when diseases occur and how they are transmitted in populations.

Centers for Disease Control and Prevention (CDC)

- Collect and analyze epidemiological information in the U.S.
- Publishes Morbidity and Mortality Weekly Report (MMWR)
    - Morbidity - incidence of a specific notifiable disease
    - Mortality - deaths from notifiable diseases
    - Morbidity rate - number of people affected in relation to the total population in a given time period
    - Mortality rate - number of deaths from a disease in relation to the total population in a given time period

CHAPTER 13: MICROBIAL MECHANISMS OF PATHOGENICITY

How Microorganisms Enter A Host

The specific route by which a particular pathogen gains access to the body is called a portal of entry.

Portals of Entry

- Mucous membranes
    - Respiratory tract
        - Microorganisms that are inhaled with droplets of moisture and dust particles gain access to the respiratory tract
        - Most common portal of entry
    - Genitourinary tract
        - Microorganism that gain access via the genitourinary tract can enter the body through mucous membranes
    - Gastrointestinal tract
        - Microorganisms enter the gastrointestinal tract via food, water, and contaminated fingers
    - Conjunctiva
- Skin
    - Most microorganisms cannot penetrate intact skin
        - Enter hair follicles and sweat ducts
    - Some fungi infect the skin itself
- Parenteral
    - Some microorganisms can gain access to tissues by inoculation through the skin and mucous membranes in bites, injections, and other wounds

The Preferred Portal of Entry

Many microorganisms can cause infections only when they gain access through their specific portal of entry.

Numbers of Invading Microbes

Virulence can be expressed as LD50 (lethal dose for 50% of the inoculated hosts) or ID50 (infectious dose for 50% of the inoculated hosts).

Adherence

- Surface projections on a pathogen called adhesions (ligands) adhere to complementary receptors on the host cells
- Ligands can be glycoproteins or lipoproteins and are frequently associated with fimbriae
- Mannose is the most common receptor

How Bacterial Pathogens Penetrate Host Defenses

Capsules

- Some pathogens have capsules that prevent them from being phagocytized

Components of the Cell Wall

- Proteins in the cell wall can facilitate adherence or prevent a pathogen from being phagocytized
    - M protein of *Streptococcus pyogene*
- Some microbes can reproduce inside phagocytes

Enzymes

- Leukocidins destroy neutrophils and macrophages
- Hemolysins lyse red blood cells
- Local infections can be protected in a fibrin clot caused by the bacterial enzyme coagulase
- Bacteria can spread from a focal infection by different means
    - Kinases, destroy blood clots
    - Hyaluronidase, destroy a mucopolysaccharide that holds cells together
    - Collagenase, hydrolyzes connective tissue collagen

Penetration into the Host Cell Cytoskeleton

Salmonella bacteria produce invasins; proteins that cause the actin of the host cell's cytoskeleton to form a basket to carry the bacteria into the cell.

How Bacterial Pathogens Damage Host Cells

Direct Damage

- Host cells can be destroyed when pathogens metabolize and multiply inside the host cell

The Production of Toxins

- Poisonous substances produced by microorganisms are called toxins
- Toxemia refers to the presence of toxins in the blood
- Toxigenicity - ability to produce toxins
- Exotoxins are produced by bacteria (mostly gram-positive) and released into the surrounding medium
    - Exotoxins, not the bacteria, are responsible for producing disease symptoms
- Antitoxins - antibodies produced against exotoxins are called antitoxins
- Cytotoxins
    - Diphtheria toxin (inhibits protein synthesis)
    - Erythrogenic toxins (damage capillaries)
- Neurotoxins
    - Botulinum toxin (inhibits nerve transmission)
    - Tetanus toxin (inhibits nerve transmission)
- Enterotoxins - induce fluid and electrolyte loss from host cell
- Endotoxins are lipopolysaccharides (LPS), the lipid A component of the wall of gram-negative bacteria
    - Bacterial cell death, antibiotics, and antibodies may cause the release of endotoxins

- - Endotoxins cause fever (by inducing the release of interleukin-1) and shock (because of a TNF-induced decrease in blood pressure)
    - TNF (tumor necrosis factor) causes edema by damaging capillaries
    - LPS stimulates release of NO from macrophages, which causes vasodilation
  - Endotoxins allow bacteria to cross the blood-brain barrier
  - Limulus amoebocyte lysate (LAL) assay is used to detect endotoxins in drugs and on medical devices

Plasmids, Lysogeney, and Pathogenicity

Plasmids may carry genes for antibiotic resistance, toxins, capsules and fimbriae. Lysogenic conversion can result in bacteria with virulence factors, such as toxins or capsules.

Pathogenic Properties of Nonbacterial Microorganisms

Viruses

- Viruses avoid the host's immune response by growing inside cells
- Viruses gain access to host cells by attachment sites for receptors on the host cell
- Visible signs of viral infections are called cytopathic effects (CPE)
- Some viruses cause cytocidal effects (cell death), and others cause noncytocidal effects (damage but not death)
- Cytopathic effects include the stopping of mitosis, lysis, the formation of inclusion bodies, cell fusion, antigenic changes, chromosomal changes, and transformation

Fungi, Protozoa, Helminths and Algae

- Fungal infections cause damage by:
  - Capsules
  - Toxins
  - Allergic responses

- Protozoan and helminths cause damage to host tissue by:
    - Direct cell damage
    - Metabolic waste products
    - Blocking lymph flow
    - Taking nutrients
    - Allergic reactions
    - Some protozoa change their surface antigens while growing in a host so that the host's antibodies don't kill the protozoa
- Some algae produce neurotoxins that cause paralysis when ingested by humans

CHAPTER 14: INNATE AND PASSIVE IMMUNITY

Overview of Immunity

Principal biologic role of the immune system is to protect us from microbial infection.

- Relatively new field of study – just over 100 years
- Organs and tissues of the immune system
    - Blood
    - Lymphatic system
    - Primary lymphoid tissue (bone marrow, thymus)
    - Secondary lymphoid (spleen, mucosa-associated lymphoid tissue)

Elements of Host Protective Responses

- Immunology is the study of the components and processes of the immune system
- Immune system distinguishes foreign substances from self-structures
- Vertebrates possess two systems
    - Innate (non-specific): First line of defense
        - Advantages
            - Always working
            - Rapid response
            - No need for previous exposure (nonspecific)
        - Disadvantages
            - Can be overwhelmed
            - Lacks memory

- Adaptive (specific): Second line of defense
    - Advantages
        - Intense response
        - Protects better on second exposure
    - Disadvantage
        - Not "on" until exposure to foreign invader (specific)
        - Slow response, can take days to weeks to rise to full strength
- These two systems work together to defend the body against foreign agents and cancerous cells

Innate immune defenses

- Found in all multicellular organism
- Provide a first line of defense against microbes
- Usually recognize biochemical differences between microbes and host cells
- While microbes can be recognized as "foreign," this system can't discern the precise identity of the microbe
    - Simply responds to an entire group of similar microbes in the same manner
        - Therefore "nonspecific" in the nature of its responses

Adaptive immune defense

- Found only in vertebrates
- Works with innate responses to achieve a stronger level of defense
    - Recognizes specific pathogen rather than broad classes of microbes
    - Response is mediated by molecules that bind to specific pathogens
- After initial exposure, it retains memory of the response used and can initiate it more quickly (and more effectively) upon re-exposure

Mechanisms of Innate Immunity

- Anatomic Barriers
  - Skin, mucous membrane, mucus
- Chemical Barriers and Physiological Variables
  - Bile, pH, temperature, iron-binding proteins
- Clearance Mechanism
  - Phagocytosis
- Enzymatic proteins
  - Lysozyme, complement
- Chemical Defenses
  - Superoxide, nitric oxide
- Antimicrobial peptides
  - Defensins

Barriers to Infection

Skin

- Generally inhospitable to foreign microbes
  - Cool, dry, acidic (pH 5.0)
  - Dead layer of cells on top
  - Layer of antimicrobial oil (sebum) lies on top of this layer
  - Sweat secretions can also provide an antimicrobial barrier
- Normal flora microbes colonize the skin and outcompete potential invaders

Mucosal Membranes

- Interior surfaces coated with wet mucus
- Moved along the surface to prevent microbe attachment

- Contain antimicrobial molecules
    - Defense proteins
    - Lysozyme, an enzyme that damages the cell wall of bacteria by catalyzing hydrolysis of the 1,4-β-linkages between NAG and NAM residues in peptidoglycan
        - Lysozyme is present in various secretions including tears, saliva, and mucus
    - Lactoferrin, a protein with bactericidal and iron-binding properties
- Competitive exclusion by normal flora

Iron as a Limiting Element

- Keeping iron for our cells is important for normal function
- Microbes need iron to grow
- Several cell types make molecules to keep iron away from microbes

The Inflammatory Response

Inflammation is an important, early physiologic response to microbial invasion and damage.

- Triggered by release of proinflammatory molecules (such as histamine and cytokines) from local cells
- Pus – neutrophil accumulation due to inflammation
- Consequences of local inflammation
    - Vasodilation
    - Increase in vessel permeability
    - Extravasation

- Consequences of systemic inflammation
    - If inflammation gets out of hand, it can be damaging
    - Can happen when microbes or their products get into the bloodstream
        - Septic shock - widespread presence of bacteria in the body induces system-wide inflammation
        - Toxic shock - overstimulation of immune responses by bacterial exotoxins in the blood
        - Death rates of 30-50% are not uncommon when septic or toxic shock sets in
- IL-1 and TNF-α cytokines are potent mediators of systemic inflammation induced by microbes or their products
    - These bring in immune cells (e.g. phagocytes) to fight the infection
    - Cytokines are often used to communicate the status of an infection
        - Producing fever (via IL-1, IL-6, TNF-α)
        - Enhancing inflammation (via IL-8)
        - Stimulating further immune responses

Molecules of the Innate System

- Pathogen-associated molecular patterns are used to sense or detect the presence of pathogens
- Toll-like receptors (TLRs)
    - A form of pathogen recognizing receptors found in vertebrates/invertebrates
    - Transmembrane proteins that recognize PAMP ligands and trigger an internal signaling reaction in self cells
    - Different TLRs can detect different PAMPs
        - TLR4 binds LPS
        - TLR9 recognizes umethylated CpG dinucleotides
        - TLR3 detects double-stranded RNA (common to viruses)

- Different TLRs may produce different cytokine responses for maximum effectiveness
    - Example: TLR4 binding of LPS stimulates production of IL-1 and TNF-α
        - Cytokines initiate a strong antibacterial response
- Mannose-binding lectin (MBL) and C-reactive protein
    - These are examples of opsonizing-secreted pathogen recognizing receptors
        - MBL coats the mannose-rich surface of yeasts and bacteria
        - C-reactive protein binds to phospholipids found in bacterial and fungal plasma membranes
        - Opsonization is the coating of a microbe, enhancing destruction or uptake by other cells

Adaptive Immunity

- Adaptive immunity is found in jawed or higher vertebrates
- System has a high degree of specificity for individual foreign molecules of pathogens
- System also has the ability to "remember" previous exposures, providing long-term protection against re-exposure to the same pathogen(s)

Features of Adaptive immunity

- Pathogen specific, induced
    - Humoral activation of B cells, production of antibody
    - Cell mediated activation of T cells, generation of cytokines
        - Cell-mediated immunity involves activation of T lymphocytes and other cells capable of cytotoxic activity against foreign agents

- Advantages
  - Immune receptors and antigen
    - T cells possess the T-cell receptors (TCRs)
    - B cells possess immunoglobulin molecules
      - B-cell receptors (BCRs) when on the surface of a B cell
      - Antibody is the secreted form of the BCR
- Lymphocytes and lymphoid tissues
  - B and T cells originate in the bone marrow
    - T cells migrate in a still immature stage to the thymus to further their development
    - Bone marrow and thymus are generative lymphoid organs
  - During development, gene rearrangements produce very large number of unique TCRs and BCRs
    - Increases the chances of a reaction against pathogens
  - Once mature, lymphocytes are expelled into the peripheral blood stream
    - They migrate through lymphoid tissues distributed around the body
  - Exposure to a new infectious agent produces a primary immune response
    - This can take 7-14 days to peak
    - Produces memory lymphocytes as a result and clearing the pathogen
  - Subsequent exposure results in a memory or anamnestic response (secondary immune response)
    - This response is faster and more potent than the primary
- All nucleated cells have MHC I
  - Only antigen presenting cells (APCs) have MHC II
  - CD4+ T-cells – peptides in MHC II (exogenous/extracellular antigen)
  - CD8+ T-cells – peptides in MHC I (endogenous/intracellular antigen)

Antigen Processing

- Extracellular antigens are taken in by endocytosis
- They are broken down and presented on MHC class II molecules
    - Restricts presentation to CD4+ T cells that can bind to MHC class II structures
    - Once presented, initiates activation of helper T cells
- Intracellular antigen and the endogenous pathway
    - Proteasomes fragment intracellular antigens
        - Small peptide fragments loaded into MHC class I molecules
    - Presentation restricted to CD8+ cytotoxic cells that can bind to MHC class I structures
    - Activation of CD8+ T cells produces effector (killer) and memory cells
        - Killer cells recognize targets by their presentation of antigen epitope fragments on MHC class I
        - Killing is achieved by T cell release of perforin/granzyme, inducing apoptosis in target

Antigen-Presenting Cells

- Dendritic cells
    - Very efficient at antigen uptake
    - Not efficient killers of microbes
    - Named for many long cytoplasmic extensions that resemble nerve cell dendrites
    - Formed in bone marrow but take up residue in maturity in various tissues
- Macrophages
    - Innate functions include phagocytosis and intracellular killing
    - Professional APC functions include presentation to memory CD4+/CD8+ T-cell

- B cells
    - Critical cells for adaptive immunity
        - Produce antibody
        - Present MHC II: peptide to Cd4+ T cells

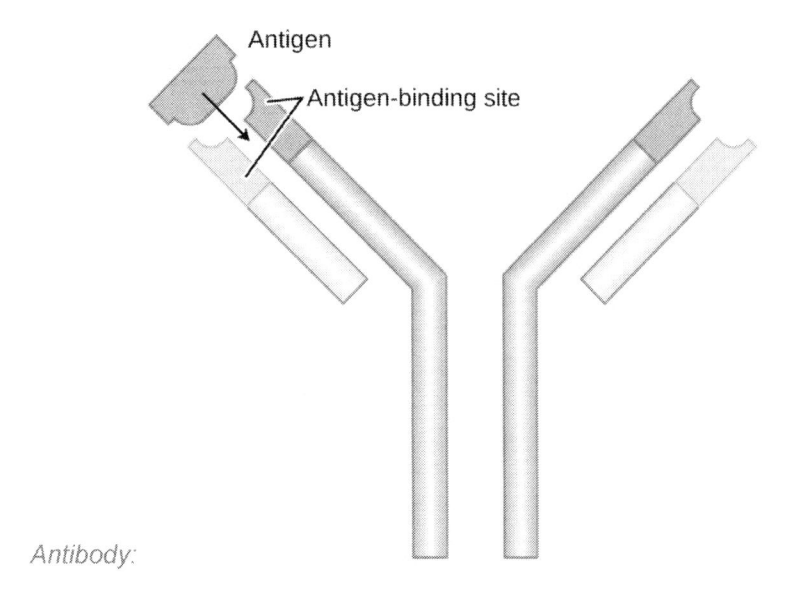

    Antibody:

    - B-cell receptors (BCRs) trap foreign antigen
        - Antigen is ingested and broken down into fragments
        - Fragments are loaded into MHC class II molecules
        - Presentation to CD4+ T cells, inducing them to secrete cytokines
        - The cytokines help the B cell to differentiate into an antibody-secreting plasma cell and/or a memory B cell
    - Most B cells need the help of T-cell cytokines to become fully activated
        - They cannot get this help without presenting antigens

Natural Killer Cells (NK)

- Useful for eliminating host cells infected with pathogens
- Not phagocytic but do make contact with target cells
    - After contact is initiated, granule components are released
        - Perforin produces a pore structure in target cell plasma membrane
        - Granzymes induce apoptosis (controlled cell suicide)

- Also useful for eliminating cancer cells
- How do they recognize an abnormal cell from a normal one?
  - Normal nucleated cells have a surface molecule known as "class I major histocompatibility complex" (MHC I)
  - NK cells recognize targets that lack this molecule
    - Virally-infected cells often turn off its expression
    - Cancer cells tend to shut down expression as well

Humoral and Cell-Mediated Immune Responses

- Cytokine secretion by T_H cells is required for proper activities of B cells and T_c cells
  - Antibodies from B cells can't directly kill a virally infected cell
  - And T_c cells wouldn't directly kill bacterial cells
- Because of the different immunity needs, there are two broad types of responses
  - Humoral immunity
    - Good against extracellular pathogens
  - Cell-mediated immunity
    - Good against intracellular pathogens

B Cells and the Production of Antibodies

- B cells produce antibodies in response to protein antigens from foreign sources
  - These responses usually require cytokine help from helper T cells
    - B cells must first bind the foreign antigen through the BCR, then ingest and process it
    - Presentation of the antigenic peptides to the TCR on the helper T cell solicits the cytokine help
    - B cell multiplies, differentiating into antibody-secreting plasma cell and memory cells

- B-cell responses to protein antigens
    - Primary responses typically produce a modest amount of antibody after 7-14 days
    - Secondary (memory) responses shift to much faster and larger-scale antibody production
        - But the ability to respond to different antigens is still retained
- Antibody production by plasma cells
    - Plasma cells can secrete 2,100+ antibodies per second
    - Terminally differentiated (cannot perform cell division)
        - All produced from proliferation/differentiation of activated naïve or memory B cells
        - Quickly undergo apoptosis (in about 2 days), but the antibodies they secrete may last for weeks in the blood
    - All antibodies from one plasma cell are identical and will bind only to a single epitope on an antigen
        - Single antigen may have many epitopes, so many different plasma cells can be formed, each with a different specificity

Protection by Antibodies

- Blocks binding of pathogens/toxins on host cells
- Fixes complement to bacterial structures, leading to lysis
- Opsonization (increases phagocytosis)
- Agglutination (clumping of antigen, increasing phagocytosis)
- Activates eosinophils, basophils, and mast cells by providing an exposed F_c region for the cells to bind to
- Antibody-dependent cell-mediated cytotoxicity (ADCC) killing of infected cells by natural killer (NK) cells

Immunoglobulin Structure and Diversity

- IgG antibody has two heavy chains and the two "arms" of the Y form the antibody binding fragments (Fab)
- "Bottom" of the Y forms the region that self-cell receptors bind to (F_c region), this is the conserved part of the antibody structure

Generation of Immunoglobulin Diversity

- Only 15,000 or so genes in humans but there are much more combinations then that that are possible

- V(D)J Recombination
    - During b-cell differentiation, there are sets of gene segments that can be recombined together to form different functional heavy and light chains
    - Different segments are used creates some diversity, but the pairing of light and heavy chains creates even more

Antibody Class-Switching

- Use of different C-region gene segments leads to different classes of antibodies
- Each class has slightly different structures and functions in immune responses
- IgM is always the first antibody in a primary response, but the complex class-switching process can change the type to make the most efficient response

Epitopes

- Most antigens have several epitopes
- Polyclonal antibodies are a heterogeneous population of antibodies
    - Each specific for one of the various epitopes on the antigen
    - Polyclonal antibodies are made by many different B cells
- Monoclonal antibodies are an entire population of antibodies that recognize the same (identical) epitope
    - They are produced by a single clone of cells

CHAPTER 15: PRACTICAL APPLICATIONS OF IMMUNOLOGY

Principles and Effects of Vaccines

Vaccines are nontoxic antigens that are injected, ingested, or inhaled to induce a specific defense response without having a person go through the disease process.

- Triggers an adaptive immune response resulting in the production of memory T and B cells specific for antigens from the pathogen
- Secondary exposure will result in a potent and immediate immune response to the specific pathogen due to the memory cells

Different Types of Vaccines

- Attenuated whole agent vaccines
    - Live but "weakened" pathogen that has been genetically modified
    - Mutant or less virulent strains of the pathogen
- Inactivated whole agent vaccines:
    - Pathogen that has been killed in some way
    - Usually by chemical treatment or by heat
    - Whole agents are generally more effective due to containing multiple antigens, but they also carry more risk of infection
- Toxoid vaccines - chemically inactivated protein exotoxins
    - Inactivated toxins are referred to as toxoids
- Subunit vaccines - a specific protein or protein fragment from pathogen
    - Purified from pathogen directly or produced as a recombinant vaccine in other organism
- Conjugated vaccines - small or non-protein antigens attached to a "carrier"
    - Necessary to enhance immune response

Herd Immunity

Herd immunity is the protection of unvaccinated people in a population where most people are vaccinated due to lessened risk of disease transmission.

- Herd immunity is lost if enough people refuse vaccination, and an outbreak can occur
    - In reality, you can never immunize 100% of the population but herd immunity means that you don't have to
    - If you can effectively reduce the number of susceptible individuals, an illness can't progress in a population effectively
- Herd immunity threshold - % of population that needs to have immunity to prevent spread of a disease
    - This value can depend on
        - Susceptibility of the population
        - How communicable the disease agent is
        - Population density
        - Vaccine efficiency
    - If herd immunity is high enough, global eradication of a disease is possible

Types of Immunization

- Passive immunization - injecting antibodies against a particular pathogen or toxin into a nonimmunized patient to provide temporary protection or treatment
- Active immunity - immunity that develops as a result of exposure to an infectious agent (natural immunization) or immunization (vaccination)

Passive Immunization

- Used to prevent disease after a known exposure (e.g. sitting on a rusty nail)
- Ameliorate symptoms
- Protect immunodeficient patients
- Block action of bacterial toxins

Vaccine Design

- Vaccines confer protection by initiating immune memory
- The ideal vaccine generates a high level of immune memory without serious side effects
- Vaccines for Fungal Infections:
    - No human vaccine for a fungal infection is currently available
    - No obvious class of antigen known to represent a class of protective antigen common to all fungi
        - Example: in bacteria you have capsule
- Vaccines for Parasitic Diseases:
    - No human vaccine for a parasitic infection is currently available
    - All parasites are complex organisms with multiple phenotypes, complicated genomes
    - Incomplete understanding of immune response

Vaccines Fail

- Failure to elicit the anticipated protective response
- Harmful side effects
- Vaccine unexpectedly makes the disease worse

New Direction

- Shot-free vaccination:
    - Nasal inoculation
    - Patch vaccine - contains many tiny prongs but the prongs do not reach the epidermal neurons so its painless
    - Edible vaccines – foods are used as vehicles for vaccination
- Genetically engineered subunit vaccines
    - Very much in progress
- Hybrid virus vaccines
- "Naked" DNA vaccines

Diagnostic Immunology

Many tests based on the interactions of antibodies and antigens have been developed to determine the presence of antibodies or antigens in a patient. These tests require both specificity and sensitivity to the antibodies. Sensitivity is the ability to recognize and bind to the antigen, specificity is the characteristic of binding only to one antigen and no others.

Monoclonal Antibodies

- Hybridomas - produced by the fusion of malignant cells and plasma cells
    - Resulting population of cells is immortal and able to produce large amounts of a specific antibody
- Uses:
    - Serologic identification
    - Prevention of tissue rejections
    - Cancer research
- Immunotoxins can be produced by combining monoclonal antibody with toxin
    - Immunotoxins are targeted to react with specific antigens

Chapter 16: The Immune System and Disorders

Hypersensitivity

Hypersensitivity is an immunological state in which the immune system "over-reacts" to foreign antigen such that the immune response itself is more harmful than the antigen.

- All types of hypersensitivity involve:
    - An adaptive immune response
        - Highly specific reactions because of T or B cells
- Prior exposure to the antigen
    - Initial exposure sensitizes the individual but doesn't result in a hypersensitive reaction
    - Hypersensitivity is only seen on secondary exposure

Types of Hypersensitivity

Hypersensitivity following secondary exposure to antigen comes in 4 basic forms:

- Type I: allergic reactions ("immediate" hypersensitivity)
    - IgE mediated and very rapid
- Type II: cytotoxic reactions
    - Cell damage due to complement activation via IgM or IgG
- Type III: immune complex reactions
    - Cell damage due to excess antibody/antigen complexes
- Type IV: delayed cell-mediated reactions
    - Cell damage involving T cells & macrophages

Note: Types I-III are all antibody-mediated, Type IV is not

Type I: Allergic Reactions

- Allergic (anaphylactic) reactions involve the activation of mast cells or basophils through the binding of antigen to IgE on the cell surface
    - Mast cells & basophils have IgE receptors that bind the constant region of IgE antibody
    - "Cross-linking" of IgE molecules on the cell surface by binding to antigen triggers the release of "mediators"
        - Mediators are histamine, prostaglandins, and leukotrienes
- Release of these mediators causes redness, swelling, itching, mucus, etc.
    - Most allergic reactions are local
        - Itching, redness, hives in the skin, mucus, sneezing
        - Usually due to inhaled or ingested antigens
    - Systemic allergic reactions can be lethal
        - Severe loss of blood pressure, breathing difficulty (anaphylactic shock)
        - Usually due to animal venoms or certain foods
        - Epinephrine can "shut down" the allergic reaction
- Some common allergens
    - Foods (e.g., corn, eggs, nuts, peanuts)
    - Dust mites, the allergen is actually dust mite feces
- Managing Allergic Reactions
    - Avoidance
        - Avoiding contact with allergen is by far the safest and most effective way of managing allergies but not can't always be insured
    - Medications
        - Antihistamines
            - Drugs that block histamine receptors on target cells
            - Histamine is still released but has little effect

- Epinephrine (aka adrenalin)
  - Necessary to halt systemic anaphylaxis
- Desensitization
  - Antigen injection protocol to induce tolerance

Type II: Cytotoxic Reactions

- Type II cytotoxic reactions involve destruction of cells bound by IgG or IgM antibodies via the activation of complement
- Most commonly observed with blood transfusions
  - Reaction to ABO blood antigens
  - Reaction to Rh antigen
- Can occur via the Rh antigen in newborns
  - Requires Rh⁻ mother and Rh⁺ child
  - Rh⁻ mother produces anti-Rh⁺ IgG following birth
  - Subsequent Rh⁺ children are vulnerable

The Rh Blood Cell Antigen

- Rh antigen is a polysaccharide on red blood cells
- Rh⁻ mother produces antibodies during birth of 1st Rh⁺ child, which can later harm Rh⁺ children

The ABO Blood Antigens

- A or B type polysaccharide antigens on surface of red blood cells
- Individuals lacking enzymes producing A or B are type O

ABO mediated Cytotoxicity

- Blood type "O" individuals
  - Do not produce type A or type B antigens
  - Produce antibodies to type A and B antigens and thus will lyse type A, B or AB red blood cells via complement

- Blood type "A" individuals (tolerate blood types A & O)
    - Produce only type A antigens
    - Tolerant to type A antigen, antibodies to B antigen
- Blood type "B" individuals (tolerate blood types B & O)
    - Tolerant to type B antigen, antibodies to A antigen
- Blood type "AB" individuals (tolerate all blood types)
    - Tolerant to both A & B antigens

Drug-induced Type II Hypersensitivity

- Involves drugs that bind to the surface of cells or platelets
- Drug functions as a hapten which in conjunction with cell can stimulate humoral immunity
- Antibody binding triggers complement activation, lysis of cells binding the drug

Type III: Immune Complex Reactions

- Caused by high levels of antigen-antibody complexes that are not cleared efficiently by phagocytes and tend to deposit in certain tissues
    - Blood vessel endothelium in kidneys and lungs
    - Joints
- Can result in local cell damage by
    - Complement activation
    - Attraction of phagocytes, other cells involved in inflammation (e.g., neutrophils)
- Antigen - antibody complexes trapped in endothelium
- Inflammatory response damages blood vessel walls

Type IV: Delayed Hypersensitivity

- Delayed cell-mediated hypersensitivity takes 1 or 2 days to appear and involves the action of T cells & macrophages, not antibodies:
    - Proteins from foreign antigen induce T_{H1} response
- Secondary exposure results in the activation of memory T_{H1} cells which attract monocytes to area

- Monocytes activated to become macrophages
- Macrophages release toxic factors to destroy all cells in the immediate area (no specificity in what is destroyed)

Infection Allergy

- Type of delayed cell-mediated hypersensitivity resulting from infection with an intracellular bacterial pathogen
    - Tc cell-mediated reaction, not an IgE based allergy
- Basis of the tuberculin test
    - Previous exposure to *Mycobacterium tuberculosis* gives a positive test result

Contact Dermatitis

- Certain substances act as haptens in combination with skin proteins
- Activates a potent T cell mediated response upon secondary exposure (e.g., poison ivy)

Autoimmunity

Autoimmunity refers to the generation of an immune response to self-antigens.

- Normally the body prevents such reactions
    - T cells with receptors that bind self-antigens are eliminated (or rendered anergic) in the thymus
        - Anergic – non-reactive or non-responsive
    - B cells with antibodies that bind self-antigens are eliminated or rendered anergic in the bone marrow or even in the periphery
- In rare cases, T and/or B cells that recognize self-antigens survive & are activated

Generating Autoimmunity

Some factors thought to trigger autoimmunity are:

- Genetic factors
    - Certain HLA (human MHC class I) alleles are associated with particular autoimmune diseases

- Foreign antigens that mimic self-antigens
    - Peptide antigens from certain viral and bacterial pathogens are very similar to certain self-peptides
    - Once an immune response is generated to pathogen, these T and B cells continue to respond to tissues expressing the similar self-peptide

Common Autoimmune Diseases

- Lupus
    - Antibodies to self-including DNA and histone proteins
- Rheumatoid Arthritis
    - Immune response to self-antigens in synovial membranes of joints
- Type I Diabetes
    - Immune response to self-antigens in pancreatic β cells (insulin-producing cells)
- Multiple Sclerosis
    - Immune response to myelin basic protein in Schwann cells (form myelin sheath of neurons)

Transplant Rejection

Transplants & MHC molecules

- Transplanted organs and tissues are rejected as foreign by the immune system due to the presence of non-self MHC class I molecules:
    - Human MHC class I molecules are referred to as the HLA (human leukocyte antigen) complex
    - There are 3 HLA genes resulting in up to 6 different HLA proteins per individual
    - There are many different HLA alleles in the human population, so each person's HLA makeup is unique
    - Close relatives are much more likely to have similar HLA antigens to recipient than non-relatives

Transplant Cells Killed by Recipient Immune Response

- Recipient has no tolerance to donor MHC
    - Recipient T cells that bind strongly to donor MHC molecules with peptide will be activated
        - Donor cells with foreign MHC class I
        - Donor APCs with foreign MHC class II
    - MHC presentation of foreign donor MHC peptides
- This leads to:
    - Activated CTLs that attack & kill donor cells
    - Activated B cells producing donor MHC-specific antibodies
        - Antibody mediated cytotoxicity toward donor cells

Identifying Donor by Tissue Typing

- Antibodies specific for particular MHC class I molecules are added to donor test cells in vitro
- Complement lysis occurs if test cells express that MHC class I molecule
- Identifying class I types facilitates finding the best matched donor

Protecting a Transplant

- Immunosuppression - drugs such as cyclosporine are given to the recipient to suppress the adaptive immune response
    - Humoral immunity is not suppressed so antibodies to donor MHC molecules are still produced
    - Some newer drugs are capable of repressing both the cellular and humoral immune responses
        - Normal immune system is impaired so there is greater risk of infection

Chapter 17: Microbial Diseases

Overview of Staphylococcus Species

- Ubiquitous human pathogens
- Most common cause of infections in hospitalized patients
- Cause wide spectrum of disease possibilities
- Difficult to treat and eradicate due to widespread antibiotic resistance
- Gram-positive cocci
- Grow in a pattern resembling a cluster of grapes
- Staphyloccoci are hearty bacteria:
    - Survive months at 25°C or 4°C in dried pus/sputum
    - Survive 60°C for 15-30 minutes
    - Survive and grow in 10% NaCl
    - Resistant to some disinfectants
- Three major species associated with human disease
    - S. aureus - major pathogen in many suppurative diseases and toxin associated diseases
    - S. epidermitis - opportunistic pathogen (hospital-acquired infections)
    - S. saprophyticus - urinary tract infections; opportunistic infections

Sites of Colonization

- Skin and anterior nares of humans (most common site for S. aureus)
- Infants are colonized with various species within hours after birth

Physiology and Structure

Polysaccharide capsule protects the bacteria by inhibiting phagocytosis (~75 % of all S. aureus strains produce a capsule). It is not produced by all Staphylococci.

S. aureus

Physiology and Structure

- Protein A - protein found in S. aureus
    - Binds to the Fc portion of the antibody, so that this portion of the antibody cannot bind receptors on phagocytic cells
    - Prevents opsoniziation
- MSCRAMMs: Surface proteins of S. aureus that function as adhesins binding to extracellular matrix proteins such as collagen and fibronectin
    - MSCRAMMs (Microbial Surface Components Recognizing Adhesive Matrix Molecules)
    - Roles of MSCRAMMS
        - Cover bacteria with components of blood or tissue so that they can hide from immune system
        - Attachment to tissue (or plastics)

Pathogenesis of Staphylococcus Species

- Some toxins produced by *S. aureus* are superantigens
    - Example: Staphylococcal enterotoxin, a TSST - Toxic Shock Syndrome Toxin
    - Superantigens - proteins that force an association between MHC complexes on antigen-presenting cells and the T-cell receptor

Enterotoxins

- Toxins that act specifically on the Gastrointestinal mucosa
- Heat stable toxins
    - Resistant to gastric and jejunal enzymes
- Produced by 30-50% of all *S. aureus* strains
- Superantigens
- Responsible for symptoms of staphylococcal associated food poisoning
    - Toxin is ingested with the food
    - Stimulates the vagus nerve endings, which control the vomiting reflex

- Symptoms:
    - Begin ~ 4hrs after ingestion
    - Projectile vomiting, abdominal pain, retching
    - In more severe cases: headache, muscle cramping, changes in BP and pulse rate
        - Death is very rare
    - Symptoms subside in one or two days
    - It is a toxinosis not a bacterial infection
        - No antibiotics are required for treatment

Toxic Shock Syndrome Toxin

- Encoded by the tst gene and responsible for symptoms of Toxic Shock Syndrome
- TSST is a superantigen - stimulates a large production of cytokines
- TSST penetrates the mucosal barriers leaving the site of infection and becomes systemic
- Syndrome is associated with:
    - Extended use of hyperabsorbent tampons
    - Cosmetic surgery patients-nose reconstruction (packing material)
    - Systemic staphylococcal disease
- Symptoms:
    - Rapid onset with vomiting, high fever
    - Rapid drop in blood pressure
    - Watery diarrhea, headache, sore throat, and muscle aches
    - Within 24 hours sunburn-like rash appear, broken blood vessels on skin
    - Bloodshot eyes, redness under eyelids, redness in mouth and vagina
    - Changes in mental status; multi-organ failure
    - In patients that survive, the rash peels off, after 10-14 days

Cytolytic or Membrane-damaging Toxins

- Example: *S. aureus* alpha (α) toxin
- Secreted toxins that either form pores or hydrolyze host membrane lipids
  - Pore-forming toxins are a common in bacterial pathogens
- Amount of toxins produced can vary from strain to strain

Exfoliative Toxins

- Toxins: Exfoliative Toxin A (ETA) and Exfoliative Toxin B (ETB)
- These secreted toxins are directly responsible for SSSS (Staphylococcal Scalded Skin Syndrome)
- Toxins target a protein found only on the surface of the skin - Desmoglein-1 (Dsg-1)
  - Role of Dsg-1: To maintain keratinocyte cell-cell adhesion
  - Degradation of Dsg-1 leads to separation of skin keratinoctyes
  - Dsg-1 is required for formation of desmosomes
- Symptoms:
  - Starts with fever and widespread redness
  - Within 24-48 hours fluid-filled blisters form
  - Skin looks like a burn or scald

Cutaneous Infections

- Local pyogenic infections include:
  - Impetigo - superficial infection on face and legs
  - Folliculitis - pyogenic infection in hair follicles
  - Furuncles (boil) - painful red nodule with yellow/necrotic center
    - An extension of folliculitis (deeper infection of hair follicle)
  - Carbuncles (abscess) - occur when furuncles coalesce and extend to the deeper subcutaneous tissue
    - Patients often have fever and chills indicating systemic spread of staph via bacteremia to other tissues

Other Clinical Diseases

- Bacteremia
    - *S. aureus* is a common cause of bacteremia often spreading from a skin infection
- Endocarditis
    - Prolonged bacteremia results in dissemination to heart
    - Mortality rate of ~ 50%
- Septic Arthritis
    - Infection of one or more joints
    - Bacteria can be identified by culturing synovial fluid
    - Most commonly affects a single joint
- Osteomyelitis:
    - Infection of the bone or bone marrow

Antibiotic Resistance of Staphylococci

- *S. aureus* and *S. epidermidis* have become notorious for resistance to many antimicrobial compounds
- MRSA (Methicillin-Resistant S.Aureues)
    - MRSA strains are also resistant to tetracyclines, macrolides, lincosamides, fluoroquinolones, aminoglycosides, etc.
- Methicillin
    - β-lactam antibiotic that is resistant to cleavage by β-lactamases
    - Methicillin replaced penicillin which is easily cleaved by β-lactamase
    - Methicillin is no longer used for treatment and has been replaced with oxacillin
        - ORSA (Oxacillin-resistant *S. aureus*)
- At one time, MRSA strains were confined to hospitals
    - Now MRSA strains can be carried by healthy individuals in the community

- New sub-categories:
  - Community-associated MRSA (CA-MRSA)
  - Health Care-associated MRSA (HA-MRSA)
- MRSA are now becoming resistant to vancomycin
  - Vancomycin used to be considered a last resort treatment
    - VRSA - Vancomycin resistant S. aureus
    - MVRSA - Methicillin & Vancomyin resistant S. aureus

Overview of Streptococcal Species

- Gram-positive cocci
- Facultative anaerobes, aerotolerant
- Non-motile
- Very hardy bacteria
- Wide spectrum of disease
- Grow in short chains (clinical specimens) and longer chains (liquid media)
- Differentiation of species within the genus is very complicated based on serology (Lancefield groupings), hemolytic patterns, and biochemical (physiologic) properties

Three Major Species Associated With Human Diseases

- *S. pyogenes*
- *S. agalactiae*
- *S. pneumoniae*

Classification Based on Hemolysis

- Alpha - incomplete lysis, green zone
- Beta - complete lysis, results in clearing
- Gamma - no zone of hemolysis

- Beta-hemolytic streptococci are further classified on the basis of surface carbohydrates antigens - referred to as the Lancefield grouping (Groups A-G) Antigen consists of cell wall carbohydrate
    - Group A contains *S. pyogenes*
    - Group B contains *S. agalactiae*

Streptococcus pyogenes (Group A)

- Beta-hemolytic (wide zone of clearance)
    - Group A streptococci
- Most pathogenic causing a variety of suppurative and nonsuppurative diseases

Virulence Factors

- Adhesins
    - Examples of surface exposed protein adhesins
        - Just like *S. aureus* - proteins are called MSCRAMMs
    - M-protein and M-like proteins - bind to extracellular host matrix proteins
        - Many different types of M-proteins (over 80 serotypes)
    - Other Functions of M-protein:
        - Binds to Fc portion of IgG and prevents phagocytosis
        - Binds to fibrinogen covering the bacteria with host fibrinogen making bacteria look-like self-cells
        - M-protein functions as both an adhesin and antiphagocytic factor
- Enzymes
    - Streptolysin O and Streptoylsin S - pore-forming hemolysins that can lyse erythrocytes, etc.
        - Responsible for β-hemolysis of Group A strep
    - C5a peptidase disrupts the complement cascade

- Surface features
    - Hyaluronic Acid Capsule:
        - Antiphagocytic
        - Prevents binding of C3b
        - Hyaluronic acid is a component of mammalian connective tissues, bacteria is disguised to look like self-cells
- Streptococcal pyrogenic exotoxins (Spes)
    - Seven known variants
        - Some are superantigens (e. . SpeA)
            - Associated with streptococcal toxic shock syndrome
        - Some are proteases (e.g. SpeB)
            - Associated with spread of bacteria through host tissue by degradation of host proteins

Clinical Disease

Clinical presentation will depend upon site of infection, strain, and associated virulence factors.

Pharyngitis

Pharyngitis is the inflammation of the throat or pharynx.

- Develops 2-4 days after exposure
- Sore throat, fever, malaise, and headache
- May or may not present with an exudate
- Cervical lymphadenopathy
- Specific diagnosis must be made by bacteriologic and serologic tests

Scarlet Fever

Scarlet fever is a complication of <u>Streptococcal pharyngitis.</u>

- Only occurs when the infecting strain produces specific pyrogenic toxins
- Presentation/Symptoms:
    - 1-2 days after pharyngitis develops, a diffuse rash appears on chest and spreads
    - Area around mouth lacks rash (circumoral pallor)
    - Strawberry tongue, blanching of skin when pressed
    - Rash disappears in 5-7 days followed by desquamation

Necrotizing Fasciitis

- Referred to as flesh-eating disease or flesh-eating bacteria (a misnomer)
- Very rare infection of the deeper layers of the skin and subcutaneous tissues
- Spreading across the fascial plane within the subcutaneous tissue
- These strains secrete many pyogenic exotoxins
- Symptoms and disease progression
    - Minor trauma - affected area becomes red, inflamed, and warm
    - Severe pain that is abnormal for the type of wound or injury
    - Flu-like symptoms (fever, vomiting, diarrhea, etc.)
    - Formation of large, dark blisters (bullae)
    - Infecting area quickly spreads
    - Shock
- Treatment
    - Antibiotics are not enough, must also be treated with aggressive surgical debridement (removal of infected tissue)
    - Amputation is often necessary

Other Suppurative Diseases

- Impetigo - skin is colonized by *S. pyogene*
    - Strains that cause pharyngitis differ from those causing impetigo

- Erysipelas - acute infection of the dermis
  - Skin is raised and inflamed most common in elderly and children
- Cellulitis - infection of dermis and deeper subcutaneous tissues

Rheumatic Fever

- Complication of *S. pyogenes* infections
  - Febrile disease develops two to three weeks following a Group A strep infection (e.g. strep throat or scarlet fever)
  - Can progress to damage to heart valves (rheumatic heart disease) and can be fatal
  - Disease is a result of an autoimmune response
- Role of M protein in rheumatic fever
  - Colonization of throat leading to sever inflammation
  - M protein enters bloodstream and elicits antibody response
  - Antibodies cross-react with heart tissue
  - Autoimmune response damages heart valves

Streptococcus agalactiae (Group B)

- Beta-hemolytic (narrow zone)
- Only species that carries group B antigen
- Consequences of surface inoculation and inhalation:
  - Pneumonia
  - Meningitis

β-hemolytic Streptococci-Treatment

- *S. pyogenes* (Group A streptococcal infections)
  - Generally, very sensitive to penicillin (e.g. penicillin G)
  - Erythromycin can be used if patient has penicillin allergies
  - Necrotizing fasciitis requires drainage, aggressive surgical debridement, and multiple antibiotics (need to inhibit growth and inhibit toxin production)

- *S. agalactiae* (Group B streptococcal infections)
    - General treatment: penicillin + aminoglycoside
    - Susceptible to penicillin (e.g. penicillin G); however, you need 10 times greater dose to clear *S. pyogenes*
- To prevent neonatal disease pregnant women are screened at 35 to 37 weeks of gestation
    - If positive, intravenous antibiotics are administered at least 4 hours before deliver

Streptococcus pneumonia

Characteristics

- Encapsulated, oval or lancet shaped
- Usually appears as diplococcus
- Alpha-hemolysis (green) on blood agar

Disease Overview

- *S. pneumoniae* is commonly found in throat and nasopharynx of healthy people
- Children are colonized as early as 6 months of age (serotypes change)
- Carriage and disease is highest during cold months
- Pneumococcal disease is most common in very old and very young
- Disease symptoms of Pneumonia
    - Rapid onset
    - High temperature
    - Shaking Chills
    - Productive cough
    - Blood in sputum
    - Pluriay (chest pain)

- Meningitis
    - Bacteria spread into CNS following bacteremia and attach to meninges
        - Meninges are membranes that cover the brain and spinal column
    - Primarily a pediatric disease
    - Bacterial meningitis is much more serious than viral meningitis
    - Difficult to diagnose until the disease has advanced to more serious stages
    - Bacterial meningitis most commonly caused by:
        - *N. meningitidis* (Meningococcal Meningitis)
        - *S. pneumoniae* (Pneumococcal meningitis)
        - *H. influnezae* type b (Hib)
    - Symptoms of pneumococcal meningitis:
        - Fever
        - Irritability
        - Stiffness of the neck
        - Drowsiness in early stages
        - Seizures and coma in late stages
    - Treatment
        - Some are still sensitive to Penicillin
        - Resistance is on the rise
        - Vaccines (anticapsular):
            - 27-valent pneumococcal polysaccharide vaccine (27 different polysaccharides)
                - Polysaccharides are T-independent antigens, stimulating B cells and not T cells
            - 7-valent pneumococcal vaccine
                - Polysaccharide + protein conjugate
                - Provides protection for children 2 years and younger

- Conjugation to protein allows for immune recognition of polysaccharide capsule

CHAPTER 18: ANTIMICROBIAL DRUGS

Overview of Antimicrobial Compounds

- Bacterial resistance to antibiotics and disinfectants could undermine major health advances (e.g. elective surgeries)
- Bactericidal - antimicrobial compounds that kill bacteria
- Bacteriostatic - antimicrobial compounds that stop or slow the growth of bacteria
- Types of antimicrobial compounds:
    - Disinfectants - compounds applied to inanimate objects
    - Antiseptics - compounds applied to skin
    - Antibiotics - compounds that can be injected or ingested

Antiseptics and Disinfectants

Antiseptics and disinfectant are chemicals that kill or inhibit the growth of bacteria and other microorganisms.

- Most are bactericidal
- Very broad in coverage – lack of specificity makes it too toxic for internal use in humans
- Tend to attack multiple targets in microbes
- Resistance to antiseptics and disinfectants
    - Poorly understood, but unfortunately it does occur

Examples of Antiseptics and Disinfectants

- Halides - chlorine (household bleach) and iodine are strong oxidants that inactivate many bacterial proteins
- Hydrogen peroxides - inactivate proteins
- Quaternary ammonium compounds - disrupt cell membranes
- Alcohols (ethanol, isopropanol) - denature proteins
- Phenols - denature proteins, disrupt cell membranes

Antibiotics

Characteristics of Antibiotics

- Low molecular weight compounds
- Kill or inhibit growth of bacteria
- Can be ingested or injected with minimal side effects
- Low toxicity but may be allergenic
    - In contrast to disinfectants/antiseptics, antibiotics are more specific
- Are naturally occurring, produced by bacteria or fungi

"Good" Antibiotics

- Few or no side effects
- Broad spectrum - active against many different types of bacteria
    - Physicians often have to treat empirically (especially in emergencies)
    - Bacterial infections often have nonspecific symptoms and it takes time to isolate and identify the bacterium responsible
    - This is a situation in which a broad spectrum antibiotic is considered good
- Problems with a "good" antibiotic
    - Broad-spectrum antibiotics not only attack the bacterium causing the infection but also your normal flora
    - Use of broad-spectrum antibiotics can select for resistant members of the normal flora that are capable of causing serious infections

Basic Sites of Antibiotic Activity

- Cell wall structure (peptidoglycan and membrane)
- Protein synthesis (30S and 50S)
- Nucleic acid synthesis (DNA replication and RNA synthesis)
- Antimetabolites

Cell Wall Synthesis Inhibitors

β-lactam Antibiotics

- General Characteristics
    - Most widely used of all antibiotics
    - Main toxicity problem is an allergic reaction to some derivatives
    - β-lactam antibiotics are bactericidal
    - Only work on rapidly growing bacteria, in the process of making new cell wall
        - β-lactam antibiotics shouldn't be given with antibiotics which slow down/inhibit protein synthesis
- How they work
    - β-lactam antibiotics inhibit the last steps in peptidoglycan synthesis
    - β-lactam antibiotics bind to and inhibit the carboxypeptidase and transpeptidase these proteins are referred to as penicillin-binding proteins (PBPs)
- Resistance mechanisms
    - Mutation of PBPs (Pencillin binding proteins)
    - Production of β-lactamases (enzymes that hydrolyze the β-lactam ring in β-lactam antibiotics (they destroy the antibiotic)
- Selected β-lactam antibiotics have been combined with β-lactamase inhibitors to increase effectiveness

Bacitracin

- Generally used in topically applied products
- Gram-negative organisms are generally resistant (cannot pass through outer membrane)
- Function by inhibiting cell wall synthesis by interfering with recycling of bactroprenol
    - Bactroprenol is the lipid carrier required for peptidoglycan synthesis

Polymyxins (Targets Cell Membranes Instead of Cell Walls)

- Insert into bacterial membranes like detergents
- Interact with LPS in the outer membranes of Gram-negative bacteria disrupting membrane structure
- Gram-positives do not have an outer membrane so they are resistant
- Generally used in topically applied products

Inhibitors of Ribosome Function

Bacterial ribosomes are an excellent target for selective toxicity because they are different in structure from eukaryotic ribosomes.

- Macrolides bind near the 50S peptidyl transferase site, blocking elongation
- Tetracyclines bind to the 16S rRNA portion of the 30S subunit, blocking tRNA anticodon binding

Inhibitors of Nucleic Acid Synthesis

- Quinolones bind to DNA gyrase
    - Causes DNA to tear itself apart during unwinding for synthesis
- Sulfonamides are a structural analog (and competitive inhibitor) for PABA, a necessary precursor for synthesis of nitrogenous bases
- Trimethoprim prevents synthesis of nitrogenous bases by inhibiting an enzyme in the folic acid pathway

Quinolones

- One of the most widely used classes of antibiotics with broad spectrum
- Inhibit bacterial DNA replication and are bactericidal

Antibiotics - More Definitions

- Antibiotic combinations - combining antibiotics to improve the outcome of treatment
    - Broadens antibacterial spectrum for empirical therapy
    - Broadens antibacterial spectrum if dealing with polymicrobial infection
    - Reduces the overall chance of the emergence of resistant organisms

- Antibiotic synergism - combination of two antibiotics that have enhanced the overall bactericidal activity compared to each compound alone
- Antibiotic antagonism - combination of antibiotics in which the activity of one antibiotic interferes with the activity of another

Mechanisms of Antibiotic Resistance

- Adaptation to selective pressures drives genetic change in microbes
- Changes in antimicrobial drug use are positively correlated with changes in prevalence of resistance.
- Increasing antibiotic treatment lengths increases resistant-microbe colonization rates
    - The more antimicrobial drug use in a facility, the more drug resistance that can be found
    - Parents with resistant strains receive antibiotics more often
        - Increases selective pressure

Molecular Mechanisms of Resistance

- Producing enzymes that modify or destroy the drug
- Altering binding targets of drugs
- Preventing drug entry into the target cell
    - Example: In Gram-negatives stop expressing a porin that is required for uptake
- Pumping the drug back out of the target cell (efflux)
    - Active efflux prevents antibiotics from reaching a high enough concentration in the cytoplasm to exert their effects

Natural Selection and Drug Resistance

- Random mutations
- Recombination
- Horizontal gene transfer mechanisms
- Most changes will have no effect on drug resistance
    - However, with continued selective pressure, rare genetic changes can quickly produce a dominant resistant microbe strain

CONCLUDING REMARKS

I hope this book has provided you tremendous value for your money and has helped you do better on your exams! If it has done both of these things, I have achieved my purpose in making this guide.

Furthermore, my goal is to create more books and guides that continue to deliver tremendous value to readers like you for little monetary costs. Thank you again for purchasing this study guide and I wish you the best on your future endeavors!

- Dr. Holden Hemsworth

Your reviews greatly help reach more students. If you found this book helpful, please leave a review on Amazon, nothing helps more than a few kind words.

MORE BOOKS BY HOLDEN HEMSWORTH

DO YOU NEED HELP WITH OTHER CLASSES?

CHECK OUT OTHER BOOKS IN THE ACE! SERIES

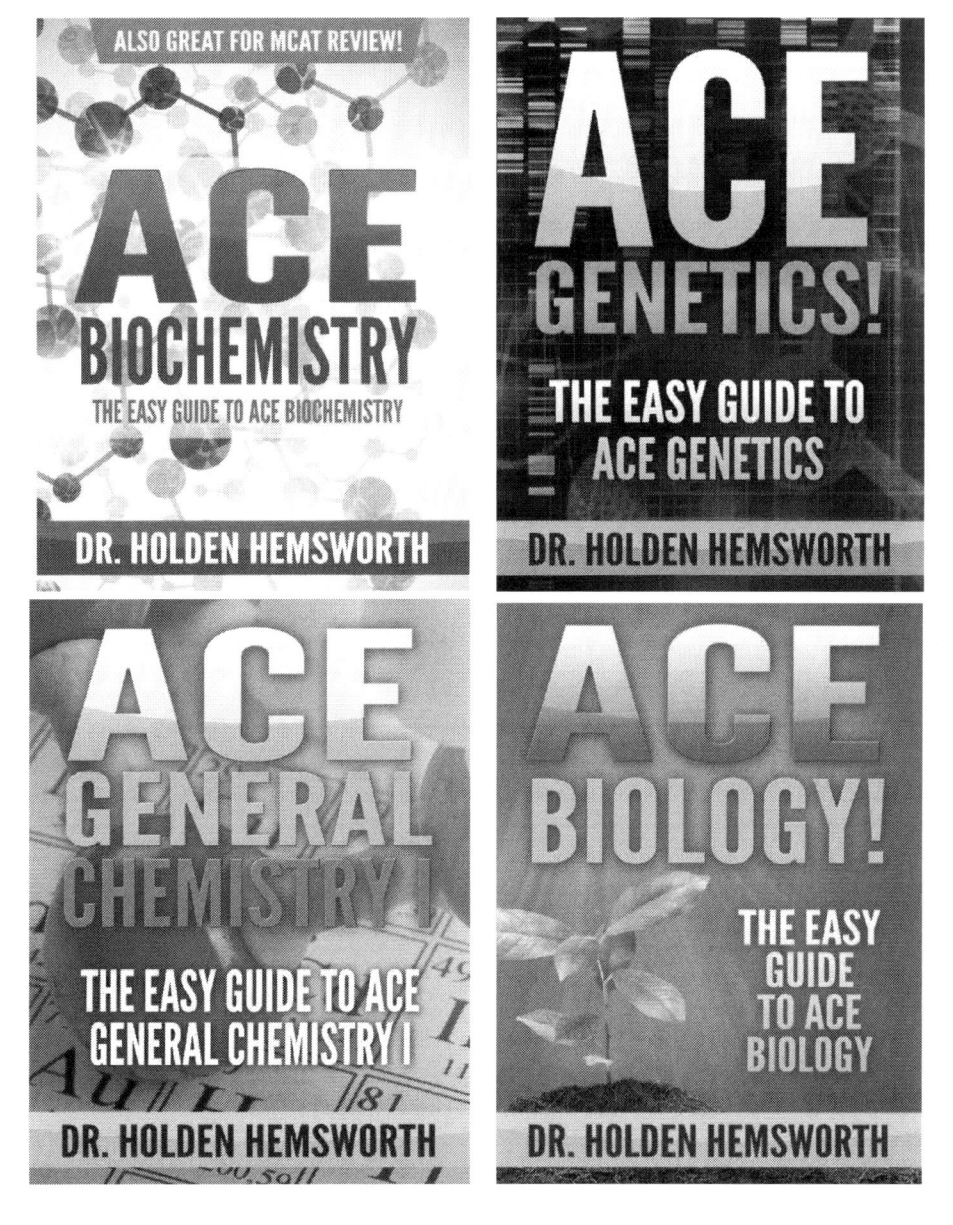

ALL BOOKS ARE LISTED ON MY AMAZON AUTHOR PAGE

MORE BOOKS COMING SOON!

Printed in Great Britain
by Amazon